CHATROOM VOYEUR

By

Donna Tracy

1stBooks - rev. 10/13/01

IN A CHATROOM YOU CAN BE ANYONE YOU WANT TO BE

TABLE OF CONTENTS

PREFACE

The world of the Internet chatroom had finally gotten the best of my curiosity when I decided to put pen to paper and try to make some sense of whom these people on my computer screen really were. Could these people be real? Could they be my friends, my neighbors, my business acquaintances, or my own family members? What about my doctor, my postman, my mechanic, my beautician? Do they all spend countless hours in a chatroom being someone else in a fantasy world where one is void of flesh and skin? Or, worse yet, could they be their real selves? What about my minister? My congressman? My brother? My sister? Does this "adult" side of the Internet world apply to them also? Then there remains the constant haunting question, "Could any of these people be my Mother or Father?" Oh, no, not MY parents—they would never be living this kind of life! And, the question we would all like to avoid, to swiftly sweep under the rug, "Could anyone of them be my husband, my wife, my son, my daughter?"

I must at this point admit to you that I am a chatroom regular. For those of you who have no idea what a regular is, it is someone who frequents a chatroom(s) and has become friends with the other regulars in the same chatrooms. For six years I have frequented several chatrooms and met some incredible people. Some I have become a close friend with, while remaining a "chatroom only" friend with others. Chatrooms have given me friendships on lonely nights and tons of laughter on silly nights. I love to attend what in chatroom terms is called a "bash", and, simply put, it is a gathering of chatroom regulars in the flesh who spend a weekend partying and getting to know one another. I have traveled across the country attending bashes and having a terrific time with great people. I highly recommend this to fun loving, adventurous people, as it is a terrific way to meet people and enjoy lifelong friendships around the world.

And, one other item I should mention about chatrooms—I met my husband in a chatroom! <GBG> (Great Big Grin).

After entering these "other" chatrooms, I must admit that my regular chatrooms are very dull in comparison. It seems we are just regular people, with husbands, wives, kids and jobs. But, after looking around at all these other chatrooms that are available to the general public, I began to wonder, "Do these regulars that I associate with have other profile names they slip into to aid them in venturing into this other world, this other side of Internet chatrooms, and become someone hiding in disguise?"

Chatrooms come in all flavors and varieties, but one night as I was scrolling through the chatrooms on the Internet, I was taken aback when I noticed that the M4M (male for male) rooms outnumbered every other type of room available for a chat. "Oh, come on, surely there are not that many gay, bi, or wannabe gay, men in this country?" Most of the men in my life are as straight as an arrow—I think. So, I started to count. 1, 2, 3, 4, 5, 6, 7, 8, 9, 10, 25, 50, 75, 100, 125, and so it went. "OK, so it's all a fantasy with these men," I was telling myself when I finally lost count and vowed to do a complete more detailed count at a later date.

Then I looked at the female rooms. There were not nearly as many female rooms as there were male rooms, but there certainly were enough to make me sit-up and start wondering all over again, "Who are these people?" "Sex—is that all there is?" "Sex available in any manner or style simply by typing?" "Sex for sale or for free?" It didn't seem to matter how the sex occurred, or with whom, as long as it occurred.

A look at several of the personal profiles and chats that were taking place in these chatrooms and I knew I had to write this book—to share these profiles with the world, as some proved just too magnificent to keep hidden from view of the world beyond the Internet. You will find a wide array of profiles in this book—some are funny, some are extremely disturbing, and some are very vulgar. But one thing is for sure; they are all genuine profiles, taken from Internet chatrooms that these people were frequenting. These profiles are on public display for all to read, respond to, and for a few, to send e-mails or IC's (instant chats) to in the hopes of acquiring a night of sex. I hope and pray that most of the profiles are nothing more than pure fantasy, but I am also sure some are extremely real, and belong to very sad, lonely, disturbed individuals.

If you find explicit sex disturbing, then please close this book, put it down, and move on, for it contains explicit language of sexual lifestyles and acts, and is not for the faint of heart.

INTRODUCTION

When one stops and considers all the worlds the Internet has brought to our fingertips, both in our business and personal lives, it is hard to believe we existed for thousands of years without this wondrous instrument. I love the Internet. I am in awe of how it was created, comparing it be an equal with any one of The Seven Wonders of the World. Could it rightfully be named "The Eighth Wonder of the World"? What type of mind was able to venture outside the realm of our average, small, pittance of a daily existence to visualize a superhighway, that in all reality has no roads, yet is capable of taking us places we have only dreamed of for years by just the click of a button? Did these gurus have the ability to see in the mind's eye just how powerful a tool they were creating? Or, did the creation keep exploding like a giant geyser, reaching higher and higher all on it's own until there was no controlling it? Did anyone really know what wondrous possibilities were being created a few years ago when the experiments started? Or, was the road plan laid long ago by an unseen element, and man was just the vehicle for the trip?

We use the Internet for shopping, travel arrangements, appointments, banking, and virtually hundreds of tasks that affect our daily life. The way our modern business world conducts business has been greatly affected, and I might add for the better, by the Internet. I believe the time has come to admit that we could never live without the Internet again, as was evident during the Y2K preparation. Plain and simple, the Internet is here to stay, and I am happy for that, as I could not live without the Internet after having allowed it to become a part of my daily life.

But, I am here to share another side of the Internet. The chatroom. There are thousands of various categories of Internet chatrooms, which can be a terrific means for locating missing persons, finding your next husband or wife, exchanging recipes, discussing your dog or cat, and for meeting people all over the world who are interested in the same things that you are. But, chatrooms have also been a tool used for sexual gratification and perversion. Sexual gratification in some lifestyles that I personally had no knowledge existed and only learned about first hand after entering the world of the "sexual" Internet.

Together we will go and explore several types of sexual lifestyles available through public chatrooms on the Internet, and we will read profiles and view actual chats of the occupants in these chatrooms that I gathered while being a "Chatroom Voyeur". Together we will try to comprehend who these people are and why they are so willing to open their lives to the public in such a sexually explicit way.

I will list various profiles from chatrooms, which I myself have found funny, interesting, sad, extremely stupid, outlandish, or brilliant in their use of the English language. These are my own observations and what I may consider funny or interesting you may find very offensive.

Some of the profiles acquired from teenagers are quite disturbing, but I felt a need to share them with parents and school personnel. Hopefully, parents will take a look at these profiles, then go and check their child's computer to see what screen names belong to their sons or daughters. If you do find any disturbing profiles that belong to your children, may you please have the wisdom, and intestinal fortitude, to discuss them with your children, as hate breeds hate if left unchecked. It is very disturbing to see in writing just how many of our young children are putting out the word to other children to hate, kill, and commit suicide, while their parents are downstairs in front of the TV completely ignorant of what their children are doing upstairs on the computer. As an eye-opener, there are thousands of children on the Internet using these types of profiles everyday. Any parent, who allows a child/teenager to have an Internet account in their own name, and not the parent's name, is asking for trouble. All parents should have the primary account name and set the blocks for their child's screen name. Just because you don't use the computer, and it belongs to the child, is no reason to give them free reign on the Internet until you feel they are mature enough to handle it.

I have spent hundreds of hours entering various chatrooms and will share complete printed conversations, which transpired in different chatrooms. Several of these conversations are extremely graphic, but bear in mind that all are real, not words pulled out of thin air or the author's mind. This is how some of your friends, neighbors, relatives, and co-workers are spending their days and nights, and the conversations they are engaging in. Now, mind you I said "your" friends, neighbors, relatives, and co-workers, because I know my friends, neighbors, relatives, and co-workers would never spend a moment of their time doing this.

All of the profiles contained in this book are printed exactly as their creator typed them, with the exception of their name being changed to protect the "innocent" (boy, is that an oxymoron). Therefore, please do not let me hear from teachers or professors about my misspelling of four letter words, incorrect sentence structure, and misuse of punctuation. These are not my words—they belong to their creators. I have also chosen to delete all links that were included in individual profiles to sexually graphic web sites. I will discuss some of these links in various parts of the book, and you may venture on your own to those web sites if you feel a need to, for I have included web links in the Appendix.

Hopefully, you will not find yourself in any of these chatrooms, but if you do please try to keep an open mind—as there are thousands of others just like you out there.

THE MEN

Donna Tracy

BI, GAY, LOOK'N

I had always been under the impression that most men worked, played sports, did yard work, and generally didn't have much idle time left after that. And, what idle time there was was used for a quick nap on the couch, fixing the car or an afternoon of sex with the wife. Obviously I was wrong, as chatrooms of interest to men seem to outnumber all other types of chatrooms by a very high percentage.

I had heard a quote a few years ago that stated, "Every four minutes a man has a sexual thought". How absurd! Are you kidding? That's ridiculous! Are you trying to tell me I have my boss' attention for approximately 3 minutes and 59 seconds, then he wanders off for a few seconds to change his train of thought to long legs and big boobs? I don't think so! What about my pharmacist, doctor, male family members? Are they all leaving my conversations within the four-minute timeframe? I pondered these questions for days and finally came to the conclusion that I may have made an error in judgement. If that statement is not correct, then who are all these men in all these hundreds of chatrooms? And, after roaming around these chatrooms one more question has surfaced, "Are most of the men on this planet either bi, gay, or married men just looking for a little extra action from either sex?" Don't go getting upset with me by thinking I am "men bashing"; we will get to the women in a later chapter. My point is, that most of the male chatrooms seem to be M4M (male for male) rooms, thus helping me to come to this conclusion; "There are several types of male rooms, but predominately male rooms are males looking for other males, or males just a "little" curious about other men." Sorry ladies, but someone had to tell you. It seems a lot of males are not too picky who they have sex with, just as long as they have sex—be it with someone or by themselves. I, of course, am speaking of sex on the Internet, talking with whomever is handy, be it male or female, which leads me also to the conclusion that men just enjoy sex however it comes—end of discussion.

How these males relate in real life with other males or females I have absolutely no idea, but I have a gut feeling they do not go around advertising for what they advertise for on the Internet, nor do they use such graphic language in their communications. Or, then again, maybe they do.

If you thought late night television was still in, think again. One Saturday afternoon I counted over 485 M4M rooms available to the public on the Internet with each room holding twenty-five people. I finally gave up my count accepting the fact that a lot of men like men.

Donna Tracy

One word of caution: I should mention at this point that not all persons in these chatrooms are gay or bi men. Some may actually be females using a male name and profile. Others may be law enforcement officials tracking pedophiles. Others could be just about anyone you may know, including would-be rapists or con artists just waiting for the prey to appear. So be careful, not stupid, as not everyone out there is really as nice as they seem or even the person they are portraying.

I have chosen to insert actual Internet chats into these pages that I have been fortunate enough, or maybe not so fortunate, to sit-in on. You may find them amusing, you may be shocked, but one thing is for sure, they are all real conversations that transpired in Internet chatrooms. Since so many of the profiles in this book contain abbreviations, and code colors and sides, i.e. red hanky, left side, there is a glossary included for your reference in the back of this book. Believe me, I had to reference many abbreviations in my journey.

Let us now begin our journey through the Bi, Gay, or Looker world of the Internet male.

Here are examples of gay men already in a relationship, but who are looking for a third partner to join their union. These types of profiles are very common among the gay community, and I might add, common in the straight couple community—but more on that later. There are many profiles requesting a third party to join in their union, but also to become a full-time or part-time houseboy in the union. In discussing this union with gays on-line, it was stated fairly often that if one joins a union as a houseboy then all of his living expenses will be paid for. He is not expected to work outside of the home, only to see to the other two partners' every wish, along with the cooking, cleaning, and laundry. And, let's not forget about the nighttime chores. He is expected to join the other two partners in bed whenever summoned.

NAME: 2 Guys, **CITY/STATE**: NY, **STATS**: 30 & 35, **DESCRIPTION**: partners but like 3sums, f*cking butts of hot guys 18-33 and smooth total tops…serving physically fit guys…no fats, olds, or balds, life is better on top.

NAME: We live in OHIO! **CITY/STATE**: Midwest, **DESCRIPTION**: Gay couple since '87: 49 & 38 Bob: 6'01", 175,br/bl & Paul: 5'10", 135, br/bl. Healthy & happy, but looking for a GM who is a submissive bottom for fun & friendship &/or wants to be our p/t houseboy.

NAME: Lovin 2sum, **CITY/STATE**: Burbank, CA, **DESCRIPTION**: very happy, professional 2sum looking for a 3rd to complete our household. Must be a pet lover as we have 3 dogs and 2 cats. Please be submissive for either top or bottom, and be into role playing. Age, 28-40.

Below is our first Chatroom Chat in a Male4Male room environment. I have found the rooms of the gay male to be extremely graphic in sexual nature when it comes to their wanting a sexual experience for the day, afternoon, or night. I myself am not sure how one continues to type while engaging in such graphic conversation, but obviously it can be done. This type of conversation takes place 24 hours a day, as you will see by other Chatroom Chats in this chapter.

A CHATROOM CHAT:

Tampa:	Lay love u
Boater:	Tampa,, you left it so nice and
Tampa:	great ass
Boater:	hot
Tampa:	yes
Tampa:	nice hole
SOCKO:	Come on please someone stick their cock in my tight ass!!!!
Layer:	your rippin me apart
Tail4U:	walking to socko
Tampa:	u can take it
SOCKO:	Hi Tail4U. I am already bent over
Boater:	hhmmm fucking Lay,,, slowly and,, trying to let him recover
Tail4U:	mmmmmmmmmmmmmm
Tail4U:	nice
Tramp:	suckin those balls
Layer:	my pussy ass can take it
SOCKO:	throw me a finger or two first
Tampa:	sitting taking a break
Tampa:	relaxing now
Layer:	fill me ass Boater
SOCKO:	or should you teach me how to suck your cock first?
Boater:	fucking Lay,,,, holding down his back,,,,, fucking himdeep and deliberate
Goater:	DAMN YES TRAMP MY ASS IS ON FIRE
Tampa:	load him up Lay

Tramp:	someone want to put their dick in my mouth with Goater
Layer:	oohhhh yeah fuck my ass hard
Tampa:	feed the ass
Tramp:	so i can suck two cocks at once...
CANDYMAN:	standing in front of Tramp cock in hand
Boater:	hhmmm fucking you hard,,, let me cum in your mouth Layer
Joe:	looking 4 phone mail me
Layer:	in my mouth or on my face
Goater:	OOHH YEA HRD NICE COCK
Tramp:	uuumm you are hard Candyman...
Tampa:	shoot the load
SOCKO:	Well Tail4U what is it going to be my tight ass is waiting??? mail me.
Boater:	on your face
Layer:	do it sir
Tramp:	suckin those cocks and fingerin those asses
Layer:	shoot your load on my face
CANDYMAN:	pushing my cock into Tramps mouth
Boater:	pulling my cock out,, and jerking it at Layer's face,,, here I cum Layer
Tampa:	give it to him
Goater:	PLAYIN WITH CANDYS ASS
SOCKO:	Come on somebody put it to me
CANDYMAN:	mmmmmmmmmmm
SOCKO:	NOW!!!
Layer:	mouth open, catching cum
Tramp:	play with his ass babe while i suck him clean
CANDYMAN:	mmmmmm nice mouth Tramp
Goater:	UUMMM YEA GREAT ASS TRAMP SO HARD AND FIRM
Layer:	your cum feels hot on my face as it drips down
Tramp:	pullin his cheeks apart for ya...mmmm
Boater:	oh yeah Layer,,,,all in your face,,,, hhmmm that was good
Tampa:	great

Most profiles of the gay or bi male are very explicit in their sexual preference requests, as I'm sure you have noticed by now. I did take note that many of the profiles requested a submissive/dominant type of relationship, one that requires ropes, chains, rubber, and other types of restraining devices, more so than any other type of relationship in the gay or bi male world of the Internet. Here is a sampling of submissive/dominant profiles from gay or bi males.

NAME: With you helpless does it matter?   PLEASE BE REAL!!! **CITY/STATE**: Tn, **STATS**: 29 5' 10" 210# bl/bl, **DESCRIPTION**: Male single, I like to have a guy with smooth legs, tied up, gagged, blindfolded. I like to caress his lower body while he squirms against the ropes. I like to roleplay pretending he is my hostage. I like having him tied up naked or in speedos or thong too guys ages 18-30 mostly, love smooth logs on a guy ☺ Having a guy in bondage is hot!!! NOT into cyber or phone or pain, torture sorry bondage alone is sooo hot.

NAME: Freddie, **CITY/STATE**: Chicago and SW Michigan, **STATS**: 43, 6'1", 200, 42"ch, 33"w, **DESCRIPTION**: Male, Single, looking for ltr, submissive bottom into sm, bd, leather, boots (especially Wescos), master/slave, bondage, dungeon in country with sling, cell, suspension, etc. also need cuddling, passionate kissing and love. Also enjoy writing, reading, baseball, entertaining, music, humor, utopia.   Very well hung, GWM, handsome, muscular, s submissive, masculine, professional.   Headline: "Soulmate dumped for new, better soulmate" The Onion

NAME: Slave Lover of Sorts, **STATS**: 36, 5,11 170 lbs, **DESCRIPTION:** Male LOOKING FOR A NEW MASTER TO SERV. Men, cops, leather, bikers, fireman, s&m, Longhair, FF, correctional officers, Ropes and Cuffs, construction workers.   THIS ONE- CAN RELOCATE FOR RIGHT MASTER. www.XXXX.com Sticks and Stones may break my bones, but rope and chains excite me.

NAME: Domestic Trainer, **CITY/STATE**: Will Travel, **STATS**: 33, 5' 7", 150lbs(no fat), into using and abusing, **DESCRIPTION**: Male, Prtnr-Japanese & boy-Blk-all play. Domesticating wild boys by means of domination. You: 21-40, <6', <200lbs, into being used and abused. How do you keep a "boy" from drowning in a mud pubble? Take your foot off his head!! No cyber or phone; and don't waste my time or yours if your not serious: can and will travel in an 8-12 hour radius on weekends; further can be arranged.

Here are a few various profiles that I thought might add an interesting insight into our world of the gay or bi male. They differ, but I guess in all actuality, they are all very similar.

NAME: INMATE LOCKED IN RUBBER, **STATS**: 45, Male, have Rubber Partner, but we play, **DESCRIPTION**: Confined and controlled in full rubber coverage for extended durations. Warden as well. Spending endless hours heavily restrained in rubber or just confined in rubber, but unable to get out of it by myself. Keyword: Frustration. Primarily into the headtrip and mindgame aspect. Rubber, latex, catsuits, wetsuits, hoods, gags, straitjackets, sleepsacks, boots, h2e, bardex, caths, etc. Not into pain, fists, sharps, black and blue, asphyx, k9, snuff scenes.

NAME: Mutt, **CITY/STATE**: New York city and North New Jersey ***NO X PIC*** Age: 5 dog years, 6'(on hind legs), **DESCRIPTION**: Male, looking for owner, licking bones, being bred, playing fetch, begging, doing tricks, handsome muzzle, hard, lean, sleek, smooth, smart, obedient, friendly, healthy, affectionate, devoted, open to all kinds of fun and games www.XXXX.com.

NAME: LOOK'NDADDY, **CITY/STATE**: Anderson IN, USA, **STATS**: Leo of 55, **DESCRIPTION**: Male, Looking LTR, son, boy, slave. BONDAGE, W/S, TORTURE OF ALL TYPES, COLLECTING EQUIPMENT. BEAR, 300#, rings in both t*t, brown ey/hair (shaved head), b*lls of a bull, HIV— starting a boys home for boys 18-45 (so make me want to collar u) WHIP ME, BEAT ME, MAKE ME BLEED, KINKY SEX IS ALL I NEED!!!

The following chat caught my attention when the occupants began discussing the going rate for performing in porn flicks with someone who finds new talent on the web.

It seems male escorts also advertise just as heavily as their female counterparts in chatrooms, as you can see by several profiles that follow the below Chatroom Chat.

Actually, I was amazed at how many actual blatant advertisements there were in chatrooms and on the Internet. Sex is definitely for sale in any manner one could phantom, on the Internet.

A CHATROOM CHAT:

PVT1STCL:	i liked the park in kansas city up from the dixie bell on the ball diamond with the lights
TellerNot:	hey Pvt...I had an offer from Yates...think I ought to do it?
TellerNot:	make some $$$
PVT1STCL:	use to have a KC cop pull me over and arrest me every saturday night
I'mFun:	Teller...how much is he paying these days?

TellerNot:	good bit...you know about him?
PVT1STCL:	he'd do me in the back of his patrol car
I'mFun:	Sure do
STUDLY:	thats too bad
TellerNot:	you done anything for him?
PVT1STCL:	it is not worth it Teller your better than that
I'mFun:	Nope...
TellerNot:	he offered $300...said could be more later
I'mFun:	That is not NEARLY enough...
PVT1STCL:	Teller dont do it your worth 100 times that amount
TellerNot:	just a jo scene
PVT1STCL:	and you know it
I'mFun:	Yeah...and he makes HUGE bucks off of those scenes...should be paying a hell of a lot more
I'mFun:	than $300
TellerNot:	could use the money
PVT1STCL:	i lend you some Teller
TellerNot:	would you man?
PVT1STCL:	yeah you know it
PVT1STCL:	if i dont get it in cash i know i will in trade
PVT1STCL:	call it a stud fee for a loss on taxes for breedin the dawg
TellerNot:	kewl
I'mFun:	LOL

NAME: Here I Am! Gay Escort Without Comparison. King of Overnight Appointments. www.XXXX.com **CITY/STATE:** Los Angeles and Travel. Please do not Email. **STATS:** 32 years HOT, Very Handsome Nordic Look: Blue eyes, Red Hair, 6' 176 lbs, 5% Body Fat. 44" chest, 17" arms, 23" Legs, 30" waist. **DESCRIPTION:** You have found your LEAN DREAM. You will never feel more secure or serene than when asleep in my strong arms after some UNFORGETTABLE HOT fun (and dinner?). You will most definitely be back for more. HARD abs with beautiful butt and built back and shoulders. Smooth Perfectly LEAN Body! A Midwest Transplant: EZ going, Masculine, passionate and VERY AFFECTIONATE: a "MUSCLE-STUD sweetheart". 3 college degrees, Impressive Passport, Kick Butt Rock Climber, Mountain Biker, Skier and Snowboarder (some surfing and sailing). Call me and instantly I will prove myself to be without comparison.

310/xxx-xxxx I HIGHLY recommend visiting my website www.XXXX.com

NAME: Blk&Sexy, **CITY/STATE**: LA CA will travel, if I don't respond to instant mail, I'm not at PC. Email or page me 800/xxx-xxxx if you want me now page me now! **STATS**: 18 yo, 5'9 155 lbs 29 w 44 chest, **DESCRIPTION**: Male, Escort an Hot Erotic Massage. Handsome and Hot Black hiv neg. Great pecs bubble butt naturally smooth, cleancut (yes I kiss too!) V-shaped Muscular checkout website www.XXXX.com otla (no attitude and fun) no pic collectors serious only. Passionate discreet (click on link below) in or out overnights.

NAME: Escort in Dallas, **CITY/STATE**: Dallas Texas, **STATS**: Male, 35, 5ft. 11in., 174lbs, 42ch, 31w, bl, br, **DESCRIPTION**: I am very masculine, good looking, smooth, hung, discreet and HIV—Escort for Bi, Married and Business Travelers. Please be in shape, very masculine and HIV. You will not be disappointed. If I am away from computer, please Email.

NAME: Lovely Gay Boys, **CITY/STATE**: Various, **STATS**: Male www.XXXX.com, **DESCRIPTION**: (Adult Video-Escort) YES LIVE NOW PRN stars live cam www.XXXX.com HAVING SEX LIVE. CHECK US OUT... Hot Boys Free Escort listings...LIVE NOW! Escorts, Videos, Watch them live. Chat Room, Hot Links, Talk on the phone live with the model of your choice. Check out the preview of the HOT Brian Jay!! Everybody's Favorite!! Hes Live now! You can tell him to play with himself or get on his knees. Anything you want to see... They will do it! www.XXXX.com Live video stream 24/7. Tell me what you want me to do! Vote for your favorite model... Purchase clothing or ANYTHING from which ever model you want. Theres a store on my site with each models article of clothing. Open Now!

A CHATROOM CHAT:

AdamA:	hi Mr. Abel walking to you
Manmade:	feels really good Soft...ber your mouth would feel even better
Mr. Abel:	Hi Adam
Honey Sweet:	kissing Mr. Abel's nipples as I rub against him
AdamA:	sit next to you
Softheart:	sucking your fat cock...Man
Manmade:	get is all wet Soft...down to the balls
AdamA:	you and sweet go a head
Honey Sweet:	swaying my ass slowly
Mr. Abel:	Grinding my leg on Honey while I greet Adam
Honey Sweet:	winking at Adam
Mr. Abel:	You can join us if you like Adam
AdamA:	love watching while i stroke

Softheart:	deepthroating that cock Manmade
Manmade:	grinding my crotch into your face as i fuck it
Honey Sweet:	stroke it in my crack if you like Adam
Honey Sweet:	Hi Adam:)
Softheart:	yeah...Manmade...fuck my mouth
AdamA:	love to baby
Honey Sweet:	me too
Mr. Abel:	I take hold of Adam and Honeys cocks
AdamA:	jerk us Mr.
Manmade:	balls hitting your chin while you swallow my cock
Mr. Abel:	sliding Adams into Honeys cracxk for him
I'mHere:	dropping towel, hard. looking for 2 or more to join
Honey Sweet:	ohhhhhhhh
TooTooKewl69:	cock swaying
Manmade:	feel hot...swalloe that precum drooling out
Softheart:	swallowing your cock...Manmade...my nose buried in your bush
Mr. Abel:	stroking bothe my men, slowly, milking their cocks
TooTooKewl69:	over here I'mHere
Mr. Abel:	getting their heads harder and redder
I'mHere:	moving over
AdamA:	oh so good
Honey Sweet:	kissing Adam
Manmade:	reachng down and grabbing your head. holding it still while i fuck your mouth
TooTooKewl69:	closer I'm
Mr. Abel:	taking Honeys head in my mouth rubbing his balls while I tug on Adam
Softheart:	like when you get rough Manmade
AdamA:	tongues twisting
Honey Sweet:	feeding my cock to Mr.
Manmade:	pulling my cock out and wiping it all over your face
Softheart:	tasting your precum...Man
Honey Sweet:	flexing my ass cheeks
TooTooKewl69:	grabbing I'ms cock
Manmade:	shoving my balls in your mouth
Starter:	want that ass licked
Softheart:	yeah...sucking your balls Man

11

Mr. Abel:	sliding a fnger into Honey and moving my mouth to Adam, letting him have a turn
Softheart:	licking those huge balls...Manmade
Manmade:	thats it, get them dripping wet with your spit
Honey Sweet:	pushing back on Mr. Abel's finger
AdamA:	getting so hot
Starter:	sit onmy face and get sucked
TooTooKewl69:	licking I'ms balls
Softheart:	like those balls...Manmade
AdamA:	leaking precumm
I'mHere:	keeping legs wide, balls tight
Softheart:	loaded with cum
Mr. Abel:	sliding a 2nd finger into Honey and deep throating Adam
AdamA:	yess Mr. Abel
Manmade:	pulling my balls out of your mouth...wanna taste my ass Softheart?
TooTooKewl69:	licking under yur balls
Honey Sweet:	spreading my feet further apart
DON'T WASTE IT:	walkin in ready to spread my cheecks for a big pole
Starter:	sucking on Honey Sweets cock
Softheart:	tasting your ass...Manmade
Mr. Abel:	Guiding Adam towards Honey's ass
Honey Sweet:	smiling at Starter
Manmade:	spreading my ass wide open for your tongue
Hard4UAlways:	hi starter suk it good for me
Starter:	yeah Honey im eating ur cock
Starter:	pump my mouth
Mr. Abel:	Pushing Adam into Honeys asshole and getting behind the, m
Honey Sweet:	feels nice Starter
Manmade:	bury your face in there and get your tongue deep up there
Hard4UAlways:	fuck him good Honey
Softheart:	tasting that hot ass Manmade
Starter:	licking ur piss slit
Missle:	Walks in confused
Missle:	hmm this looks interesting
Starter:	shove it in
BEACHBOY:	lean aggressive muscleboy sweaty and horney from my workout

Mr. Abel:	I slide up gehoind you, my hard on leaving trails of precum on your ass
Softheart:	my face buried in your ass...htf
Honey Sweet:	running hands over your face Starter
SadSam23:	slowly stroking, looking arounf
Manmade:	pushing you down and practically sitting on your face
DON'T WASTE IT:	mmm my tight hole needs to ride big pole
Starter:	balls bouncing on my chin
Softheart:	yeah...use me Manmade

Following are profiles of SheMales. A SheMale becomes a female in the "feminine" sense by having breast implants, hormone treatments, and dressing and acting like a female, while remaining a male in the genital area. The male who actually becomes a full blown SheMale goes through a series of hormone shots and breast implants to achieve real feminine characteristics. One would never suspect that the female who is in the picture, or who you are having dinner with is actually equipped with male plumbing. Many a male has been quite surprised when he takes the cute blonde he has met in a bar after a night of drinking to a hotel room, and the truth is unveiled. I personally found the photos of SheMales that are on the web to be, how would one put it, "Different" to say the least?

NAME: Wild Kitten - ~~~ VERY EXOTIC SHEMALE~~~ {{Kisses&Hugs}} **CITY/STATE**: West Hollywood, CA, **STATS**: Aquarius~'75, Doctor, Researcher/Editor & Part-time Ramp Model Single untilYOU FEED ME w/ LOVE, **DESCRIPTION:** Pleasing and Making Man HAPPY and SATISFIED To it's FULLEST. I LOVE Ita/wht Str8Dominant 25-40 yrs. Old Muscular goodlooking MAN. "Whatever You Need, Want, Fantasize & Dream & I WILL FULFILL www.XXXX.com

NAME: Flower, **CITY/STATE**: Western USA, **STATS**: 53 years young...if this bothers or offends you...please move on cause it is not a topic for debate. Single...where is Mr Right? **DESCRIPTION**: Movies, Museums, White Water Rafting, Classical Music, Travel. Who cares...NO female pix please...NO crossdressers...Love Bodybuilders, but they usually don't like me, but a gal can dream! Make enough to pay the bills and travel when I can.

NAME: Exotica, **CITY/STATE**: Chicago, IL, **STATS**: Single, but hopeful, **DESCRIPTION**: If you don't have any pictures, don't waste your time to mail me! I'm not being rude, I just like to be honest and straightforward. I only date men between 18-28 years of age. And, they have to be cute and in shape. I'm sorry if I'm picky and superficial, but everyone has their own taste/type. And sure enough, I do too. How do I look you ask? Well, I'm Chinese, tanned, 5'10', weigh 135 lbs, black hair and brown eyes. I make clothes for my friends and myself. Don't be shy to say hi! I consider myself a chick with a dick. If you are looking for a full-time girl, I'm not quite there yet (started hormones). By 2001, I hope I look real so I can live full-time as a "girl" with a dick.

A CHATROOM CHAT:

BillyBoy:	any young tuff tops here?
Desperate For It:	HOT MWM want to get someone off NOW
Desperate For It:	Into ANYTHING that YOU like (no shit blood or minors)
Lookin2Be:	any top who likes it kinky? mail me. will do ANYTHING
Desperate For It:	MAIL me your fantasy I want to get you off
Desperate For It:	can go from str MAN stuff to hose and panties
Desperate For It:	love getting into my partners fantasy MAIL ME
Desperate For It:	Sorry MAIL was blocked
SleeyMan:	18 Jock with self pic to trade and hot phone looking for a bro or a daddy mail me
Desperate For It:	MAIL me for GREAT cyber YOUR way
LUV GAMS:	Looking for built guy like me (with pic) to make this top a bottom!
Desperate For It:	NOONE wants to play?\
Gone Now:	Hot stud for hot phone with hot Dad/top...
LUV GAMS:	Sucker for a guys legs.
Desperate For It:	WHO WANTS TO GET OFF?
Desperate For It:	IC me with YOUR fantasy LETS DO IT!

Now we find some very interesting profiles in the sense that if you have a foot fetish, then these are for you. It amazed me how many men considered their feet a sexual object. I don't think I could quite equate men's feet with men's biceps, but then I'm not a gay or bi male with a foot fetish. I did check out these guys web sites, and decided their feet were nothing to write home about, at least for me.

NAME: FeetInMyFace, **CITY/STATE:** UWS, NYC, **STATS:** 33, Male, Single, **DESCRIPTION:** Sneaks, Socks and Feet, Pits, kissing. Don't send me faceless photos please, not interested in men without heads.

NAME: TWELVESINSF, **CITY/STATE:** San Francisco, **STATS:** 42 yo top, involved, 5'10" 160, **DESCRIPTION:** Scroll down for link to web site w/my feet pix. Ask about buying my used socks (white or dark). Also have a well-worn pair of Nike Airs available. My target market: masculine, in-shape guys up to age 35 with feet up to about size 11. Not interested in pix of just feet; I'm interested in the whole package. No profile, no chat. None. Get it? Enjoy having my clean size 12s licked, sucked and massaged. Also like doing the same with good looking, in shape guys up to age 35. Clean feet only, white sox a plus. Also like hair legs. www.XXXX.com

The first of the "masculine" profiles below, is probably my favorite of all the profiles I have come across in my journey through the gay male chatrooms. I can just picture this over masculine, testosterone filled, short, military guy declaring at the top of his lungs that he is straight—just got this little itch. "Ya, Sarg, you and about 20,000,000 other "straight" guys."

Some gay or bi men prefer to only look for very masculine partners— partners who are not a threat to their own masculinity. They prefer a very discreet meeting with a very masculine male, which, I suppose, enables them to remain in denial about their own sexual preference.

NAME: CALL ME SARG!!, **CITY/STATE**: Texas (that's all ya need to know), **STATS**: 27 yo, married (TO A WOMAN), 5'5" 160#, brown/brown, **DESCRIPTION**: I am straight... totally, and yeah, I have DONE IT (verycurious) HELL YEAH, and if ya got a problem with it, go cryin somewhere else (and that doesn't mean I am a homo) I have a twin bro who eats up all my time. If I want a lady, (which I LOVE) then I can get one, but sometimes there's this "itch." Military here, and don't ask "which bra nch, if I was to tell ya, I would have to kill ya, (or Dishon. Discharge) Military cut hair, great shape and clean cut (yeah, cut THERE too) If you ask to trade pics you send first. I only have a g rated pic. Best Distance: 4 FEET 8 INCHES! HOOOOAAAAA!

NAME: BULLDOG, **CITY/STATE**: Can Travel, **STATS**: 47, brown hair, ITAL, 5'8"/183, **DESCRIPTION**: Looking for discreet guys in the area—get off on super masc. Tops-Like tatoos, workboots, think belts and guys that are guys. I'm muscular humpy guy in trips to the woodshed-real good at servicing you and buddys if youre into it. SERVICING YOU AND BUDDYS IF YOURE INTO IT. BROCKTON, CAN TRAVEL. If youre a regular guy with heavy equipment, like being the boss And get off an a masc. Bottom, IC me. Blue collar gets my attention immediately. I like regular, horny, hardworking guys, straight or BI great! IM a total bottom-real good at it-youre a total top! If straps and tatoos are a turn on get back to me. I'm inked-like bikers, bluecollar, and any extemely masculine tops out there. Sit back and relax buddy-let me do the work!! All ages over 30! Very discreet-expect same!

NAME: EXECUTIVE FOR YOU, **CITY/STATE**: DALLAS TX, **STATS**: 34, 175 LBS, 6', brown hair, blue eyes, **DESCRIPTION**: Executive in dark over the calf socks, suit, tie and great shoes. Love to play with other well dressed execs; will shine shoes for the right execs while in my suit. Looking for uniform scenes with intense bondage and discipline. Looking to polish boots of highway patrolmen, military officers.   Am muscular, very good shape, "straight" appearance, very clean cut.

A CHATROOM CHAT:

Twist&Turn:	oh yeah
FRANKY:	sucking u twist
Twist&Turn:	i roll over on my back, throwing my legs up in the air
Twist&Turn:	and over my head
Horn Blower:	yeah, plowing your hot pussy hole
BLKforCyber:	dayum...look at twists hole
BLKforCyber:	wide open
Twist&Turn:	cum on...dip your cocks into it
BLKforCyber:	horn...can i fuck twist
Horn Blower:	sure think blk, bring that big thick black cock over here
Twist&Turn:	hell yeah
Horn Blower:	i wanna guide it into his pussy hole for ya
CALL4ME:	hello men... * tips hat...
BLKforCyber:	wanna drill dat puzzy
Twist&Turn:	only if you call me names while you fuckN me
BLKforCyber:	listen cunt...u like nigga dick
Twist&Turn:	fuck yeah
BLKforCyber:	take this
Twist&Turn:	give me that black donkey dick
BLKforCyber:	FUCCKKKKKKKK
Problem:	hey
Problem:	that fagg
Problem:	said
Problem:	kkk
BLKforCyber:	drillin yo puzzy
Problem:	like 5 times
Twist&Turn:	u like fucking white pussy boys?

CALL4ME:	hmmm did i walk in on somthin guys?
BLKforCyber:	yup
Problem:	no!
Problem:	just a bunch of Faggs

The following profiles are a mixture of sex for hire, off the wall sexual preferences, and some "pig" profiles. I have included other "pig" profiles in a later chapter, but felt a few should be thrown in here for your enlightenment as they belong to a very select group who find sexual satisfaction in the most disgusting sexual form known to mankind. The world of the "pig" is a real world to these men and I can honestly say, "I hope I do not know any man who would label himself a pig".

NAME: Slave for Sale, **CITY/STATE**: Los Angeles, close to Hollywood, Downtown, WeHo, east Valley, **STATS**: I'm 33, your age does not matter, **DESCRIPTION**: Master's Bidding, or???? (Escort Pig for you) Versatile, usually bottom, but... Limits are a sign of weakness, what's your pleasure, sir? Mine costs less than it should...ask for a quote (groups are great) My long tongue, your bu##; my throat, your c**k; my gullet, your pi$$ Chemical-free pig is well-trained, but still young novice. Will serve all (age, weight, ract), submits to almost anything.

NAME: Giving You Oinks, **CITY/STATE**: Orlando, FL, **STATS**: Single, 40, 6'3", 215#, 45"chest, 34"waist, brn hair/eyes, sometimes goatee, great smile and legs, **DESCRIPTION**: mostly deleted but vac pumping, leather, rubber, boots, red and yellow right and left hoods, cycle breathing breath control, toys, very chem friendly, want to try horse. Lets go to the sty!

NAME: LITTLE BOY PIG SLAVE, **CITY/STATE**: Brooklyn, NY, **STATS**: 27, 6'1, 185, WORK OUT, **DESCRIPTION**: D & D free, Sub slave for extreme in shape masters, bondg, x treme cbt, ff, ws. PIERCED NIPPLE TATOOED KINK & RAUNCH & S & M. A little pain goes along way. Much pain goes further.

NAME: THE HWY TRAVELER, **CITY/STATE**: Newbury Park, CA, Vegas, Albuq, Iowa City, MplsMN, PhxAZ, **STATS**: 67 yrs old, GWM, Single, 5'11", 175lbs, balding, hairy chest, 5.5c prosthesis (implant), Retired, **DESCRIPTION**: Kink bot (bareback or covered), ws, K9 Curious (I don't have a dog), spank me, enema curious, no pain....... HIVneg, you be also. I prefer young guys but I like all ages. If God made us both the same, one of us wouldn't be necessary.

NAME: OAKIE, **CITY/STATE**: Tulsa, OK, **STATS**: Male, **DESCRIPTION**: Seeking 24/7 Master, Group scenes, WS, CBT, h2e, bbk friendly! K9, nullo, snuff fantasies. Anything NOT vanilla, looking for some charged cum. Anyone wanna broaden my horizons?

NAME: LITTLE GUY, **CITY/STATE**: Los Angeles Area (near Disneyland), **STATS**: 5'2", bit of a belly, 43, **DESCRIPTION**: S/M from a little guy; t/t, cbt/ paddlin/spankin, wax, other fun pain as pleasure. Can also get into domination werestling/body punching-have mat, gloves, videos. Looking for true masochists. Wanna hurt for me, boy? I like to hurt, not injure!

NAME: TIGHTEND, **STATS**: Blkm, 6'2" 190# dark skin average built nice looking, **DESCRIPTION**: Hot Mouth, Deep throat, Tight A** NOT INTERESTED IN YOUR DICK PIC WITHOUT YOUR FACE. I HAVE SEEN A FEW UP CLOSE. Very Masculine A real man and I want a man. I am no Mr. America but I am average built and in good shape. I am every thing a great "Bottom" should be. I have every thing a great "Bottom" should have. I do very well, what every great "Bottom" should be able to do. If you are versital hit the X (most versitals are really bottoms.) If you are fat hit the X. If you are a hot nice looking in shapt Top lets talk. Don't wont know long skinny d*ck (they hurt) don't wont no short fat d*ck (they hurt for nothing) give me an average to above average long thick d*ck. www.XXXX.com

NAME: YOUCANSLAPMEANYTIME, **CITY/STATE**: San Francisco, **STATS**: 6'1", 188, 6"thick, cut, well built, furry, gotee, buzz cut, single and happy, **DESCRIPTION**: Slap me with your big c*ck! I like real men, Straight, Bi, Gay…If you have a BIG one and love to show it off! (don't mail me unless you want to get together NOW!) Above average…solid hard drive low hanging. So slap me with it, rub it on my face, roll your B*lls all over me, stick it in my mouth, I will stick my tongue in your H*le. "I challenge you to zzee doule!" (slap) P.S. You are welcome to squirt your load anywhere on me at anytime.

NAME: BootyUp, **CITY/STATE**: Miame, Ft. Lauderdale, Pompano Beach 3/12-3/19 Lookin 4 hot boyz—**STATS**: I'm born again when U stick it N, Male, **DESCRIPTION**: Blk B*tch, Yo Hoe or Freak on the Bottom—Your choice. You Whip it good you wed me—I love to raise the A** high in the air then slide it down on a long fat pole, that's how I message the hole. Big head makes me feel loved. The door to my "D" drive is tight, Light, sexy and sensual when hit just right… Hookin boys with my hot a** looks and Sexy body. Stretch me out and nail me to the mattress. It may take 1,2 or 3 different nails to hold me down.

A CHATROOM CHAT:

Storeminder:	good afternoon...love coming here to have sex with older men
Bo4U:	Holding my boy tight in my arms, giving that wink of approval
SLVtonight:	ass channel filled to overflowin with Bo's sperm
SLVtonight:	bo's sperm drainin out my tight boyhole
Bo4U:	and smiling
SLVtonight:	poh christ sir thank you sir
SLVtonight:	smiling kissing bo
FLYER:	cool...working my tongue up slvs thigh
Bo4U:	Would you like to lick the cum off my cock now?
SLVtonight:	feeling flyers hot tongue
FLYER:	licking the head of bo4u's dick as he pulls out of slv
SLVtonight:	yes sir let me lick you shaft while gap licks my hole
Storeminder:	45/BiMM...8 1/2 x 6 cut
Bo4U:	pushing flyer to the side, "That belongs to SLV"
SLVtonight:	let me suck it clean
SLVtonight:	maybe he can have your manjuice from my hole sir
FLYER:	okay let me eat that hot ass SLVTonight
Bo4U:	Yeah let him suck me out of your hot ass
FLYER:	can i
SLVtonight:	is it okay Bo sir?
Bo4U:	Yes
FLYER:	that tight hole is nice SLV
HrdasRock:	walking in stroking hard 7
Tequilla:	hey men
SLVtonight:	sucking bo's big cock deep and cleanin it
FLYER:	with the cum oozing out of it
Tequilla:	hard as a rock
SLVtonight:	feels great flyer, push that tongue in deeper
FLYER:	cum on over tequilla
Tequilla:	i start walkin and ooooppps
Tequilla:	towel falls off
Bo4U:	running my finger's through your thick blond hair
FLYER:	too bad

SLVtonight:	lesnin against Bo's chest
FLYER:	oh that ass is nice SLV
Tequilla:	11" stickin out!
BLPF:	m/24 walkin in towel wrapped around my waist... checkin out the action... lookin over at
SLVtonight:	yeah Flyer suck it eat it tongue-fuck it
FLYER:	slap me in the face with it Tequilla
BLPF:	Tequilla
Tequilla:	hey BLPF
BLPF:	how u doin stud
SLVtonight:	leanin against bo's chest, feelin safe
Tequilla:	ok you
Bo4U:	watchin flyer eat my cum out of my boys ass
FLYER:	slv's ass i till nice and tightr
BLPF:	doin good here
SLVtonight:	chest heaving panting breathin hatd
Bo4U:	holding slv tight against my chest
Theater Stud:	ok boys hottt here needing a huge plug
FLYER:	with my tongue working up it so slow
SLVtonight:	pushin my boyhole back against flyers tongue
MichaelM:	walking in dropping towl checking out all the hott dexy guys
FLYER:	let me taste bo's cum
Tequilla:	several nude pics here
SLVtonight:	kissing bo's chest, kissing his nipples, beginning to suck on one of them
Theater Stud:	sit back and watch stroking my hardening cock
Tights&Kisses:	walk in drop towel check the room out
Bo4U:	let's see that big one tequilla
WetnCrzy:	show me slvtongith's ass pic
SLVtonight:	sucking bo's nip while flyer tongues me deep
BLPF:	wouldnt mind seeing them tequilla
FLYER:	tasting that hot ass of slv's
Tequilla:	anyone who wants mah pic press 33
WetnCrzy:	let me in flyer
MichaelM:	33
Bo4U:	Pressing your face into my chest
FLYER:	tasting that hot cum filled hole
FLYER:	bo's dick is getting hard again
Theater Stud:	come here and kiss me someone
SLVtonight:	yeah pushing my boyhole up for flyer and others to suck and tongue it while i biury my face in

FLYER:	let me lick the head of that for you bo
SLVtonight:	my man's chest
MichaelM:	Waiting for guy to walk over
WetnCrzy:	i am bent over and available tooo
SLVtonight:	squeezin my nips and suckin bo's nips
LooknForOwner:	movin to suck bo's other nipple while I watch SLV
FLYER:	wetncrazy sit on my face
Bo4U:	Sticking my tongue in your mouth
LooknForOwner:	reachin for slv's nipples to tug on\
SLVtonight:	oh god suckin bo's tongue
FLYER:	while i eat that ass for you slv
SLVtonight:	wetn tongue-fuck me too
WetnCrzy:	on your face flyer
Theater Stud:	time to fuck boys
FLYER:	and working my way up to bo's dick
Theater Stud:	cum and get a 9" cock
LooknForOwner:	fuck me THEATER
Bo4U:	watching the guys eat my boys ass
FLYER:	cum feed me THEATER
SLVtonight:	need my boyhole cleaned, bo's cum stilkl drippin out
Tights&Kisses:	walking over to theater
WetnCrzy:	lifting my ass wet and ready for someone
Dog Bone:	i'll clean you out slv
FLYER:	i am...
Dog Bone:	~~~~~ tasting bo's sweet cum
FLYER:	you can have some to Dog Bone
SLVtonight:	still suckin bo's tongue my hands on his nips, my boybutt in the air
Theater Stud:	kiss me tights
Tights&Kisses:	turn around and bend over for him
FLYER:	get on your knees next to me
Dog Bone:	mmm -thanks gap
Tights&Kisses:	kiss theater boy deep
Dog Bone:	kneelind behing slv's sweet butt
	theater, need an ass or cock to work as well
SLVtonight:	oh god dogbone yes, wigglin my butt clampin down on your tongue with my boyhole
LooknForOwner:	fuck my ass tights?
Dog Bone:	ohhhhhh - hot ass, man
FLYER:	can i cum over tights

21

Tights&Kisses:	nah need one to eat
FLYER:	you can lick my big dick
SLVtonight:	kissing bo's nips longingly, deeply sucking on them gently, first one then the other
SLVtonight:	bubblebutt boyhole pointed in the air
Dog Bone:	pulling on slv's balls and diving in for more bo cum
Tights&Kisses:	fuck me hard theater
Dog Bone:	got 8 fat inches for you slv
SLVtonight:	is it ok Bo sir?
Aren't I Jolly:	walkin in leanin against the wall rubbin the bulge in my leather jock
FLYER:	can i suck it first Dog Bone and get it nice and wet

You have now read a sampling of the world that exists for the gay and bi males who frequent M4M Chatrooms on the Internet. This is a vast world that seems to have no end, like a black hole in space it is continuous. The Internet is inundated with web sites and chatrooms for the gay or bi male, or those who are just curious and browsing around on a Sunday evening, like myself. It contains everything anyone would ever want to know about this lifestyle, and everything one would not want to know.

But, while most occupants of the Male4Male chatrooms are extremely open to discussing their sexual preferences, and have no inhibitions when it comes to requesting sexual favors of others in the chatroom, others are extremely closed mouthed and will discuss nothing with an outsider. You will find the majority of these chatrooms packed almost twenty-four hours a day, leading one to believe that this lifestyle will be around until the end of time.

I found this world filled with love, hate, and pain between men—one which I never knew existed quite to this extent, but exist it does, as was evidenced in my journey through these chatrooms and web sites. But, even after this eye-opener of a trip, I still want to believe the majority of men are straight and enjoy the sight of a naked woman more than the sight of a naked man.

STR8 SUBMISSIVES

In comparison to the gay or bi male chatrooms, which are demanding, sexually blatant, and outright violent at times, I found the STR8 (straight) Male Submissive chatrooms to be filled with occupants wanting basic everyday chatter, even though their profiles clearly stated their chosen lifestyle preference—to be completely dominated by a woman. Their conversation ran from everyday stresses to what to cook for dinner, a far cry from the M4M rooms, who seemed unaware that food even existed. In these chatrooms I observed that most occupants would take the time to try to explain their lifestyle to anyone who was on a quest to learn, and I might add, with humor thrown in at times. They would always offer links to a web page where one could go and learn at their own pace, instead of offering a first hand lesson in the chatroom, and they could be abrupt putting anyone in their place who wanted a "quickie" lesson. This is a sincere lifestyle to these men, and they say it takes years to learn and is not something to be entered into lightly.

I cannot possibly tell you why they have chosen this lifestyle, but it is one they try to live everyday, even though it demands the woman to be completely in control of the submissive's life; to make the decisions, to inflict the punishment, to have the submissive learn to adore, respect and love her. It is a complete worship for the dominating woman from the submissive man.

This lifestyle can offer great pain to the submissive as it revolves around torture as punishment, which ranges from spankings to actual gagging with incredible suffering inflicted on the submissive. Persons in the submissive/dominant lifestyle will disagree with this statement as they believe accepting this type of treatment shows great love and trust on their part for the dominating partner.

This lifestyle belongs to all types of people from many different walks of life, but it was obvious it contains more white-collar workers than blue-collar workers. The majority of submissive men are educated professionals, with high-pressure jobs.

During my travels through the submissive male chatrooms I encountered a fourteen-year-old boy who was looking for a chatroom regular. He was professing great love for her and wanting to serve her as a submissive. In another room I found a young girl who was screaming at the occupants that she was fourteen years old and wanted to do bondage. Both left the chatrooms after being told politely, or not so politely, to move on. This gave me a wake-up call as to how many of our "children" are wandering around these sex chatrooms and gaining their sex education on the Internet instead of in the classroom or from their parents. Let me remind you here that most of the sex chatrooms contain some type of web link, or web links are inserted into profiles, and with one click of the mouse you are transported to extremely graphic, sexual photos and literature concerning the type of lifestyle these chatroom persons prefer. These web links I speak of do not require any credit card to enter them. They are free, but of course they offer you the choice to pay a membership fee for the really good photos. Believe me, what I saw for free was "good" enough.

Here are some of the profiles and chatroom chats that belong to submissive men who are, or are wanting to be, dominated by a woman.

NAME: Wil AdoreU, **CITY/STATE:** NY/NY, **STATS:** 6', 175#, Handsome, athlethic build, professional. **DESCRIPTION:** Shy heterosexual sub. B&D, C/D, and everything else life has to offer outside of D/s. If you do enslave me, you truly are the spherical one. Looking for strict but nurturing sincere Mistress for long term monogamous Dominant and submissive lifestyle, please. Would like training and conditioning from accomplished single or divorced Domina, please. Adoration, servitude, compliance, pampering, ecstasy, and real love! "To have a friend is to be a friend". "Respect is a two way street". We must speak over the phone before exchanging pics. Consistency and actions determine merit and only time will tell? Can travel or relocate.

NAME: CallMeSally, **CITY/LOCATION:** in the presence of Mistress Lilly, **STATS:** male, **DESCRIPTION:** Mistress Lilly has named me "Sally", I belong to Mistress lilly. well, i used to be tough and wild, but Mistress Lilly has taken me in and turned me into her feminine plaything; i may only type in pink letters to her and she is teaching me how to be a sweet little submissive for her pleasure. i find this an almost constant humiliation (which I secretly love, Mistress)—i am almost always blushing now <<<blush>>> i am Mistress Lilly's personal maid, in heels and stockings and a very short dress <<<blush>>> the most precious moments to time—when the night sky yields to dawn, and when the late afternoon bends before the will of twilight—are pink

NAME: Beloworabove, **CITY/STATE:** Chicago, IL, **DESCRIPTION:** Being dressed as a sissy slave and serving a beautiful mistress who can make me into the total woman I yearn to be being a sissy slave. I yearn to serve a mistress in all capacities.

NAME: Knocasinknockmearound ☺, **CITY/STATE:** Southwest, **STATS:** 33, single, **DESCRIPTION:** My previous girlfriend was a very aggressive, very rough BBW Domme who introduced me to D/s and BDSM and other related activities, and I would like to meet another for r/t relationship ☺. Very aggressive Dommes who like to humiliate and physically punish their subs welcome! Forced fems with latex, leather, rubber for the right Domme, among other thing! You never forget your first BBW Domme! Especially mine! She never told me the first session would be so violent, but, I was hers!

I have never met or been with a man who wanted me to parade him around in front of my friends. In fact, most of my male companions would go out of their way to avoid being with my friends. And, this now brings me to another thought, "Ladies, look what we may have missed all these years by not having a submissive male in the home!" He cooks, he cleans, he shops, he gives manicures and pedicures, he can probably do hair, and he will adore us! And what is required of us? A swift kick to his butt every now and then (which we have all had the urge to do a time or two to our husbands, anyway). Maybe we need to really think this one through a little harder ☺

NAME: BIG Boy, **CITY/STATE:** Boston and Las Vegas, **STATS:** 30yr, 225#, 6'5", blonde and blue, BECH 345, **DESCRIPTION:** Pleaseing and being told and spanked often. Serving a faithful mistress—cleaning, being paraded for you and your friends.

NAME: strap4me, **CITY/STATE:** Ogden, UT, **STATS:** Single, 31, **DESCRIPTION:** exhibitionist...love to strip, especially for hot dominant women...like to play with me wanna watch...or should I ask permission...thank you mistress...i got the body if you got the time...love given and getting totally nude massages...being tied up/spanked whatever lets me show off this perf body. Serving tan doms. exhibitionism with total starngers being a sub...need dom woman to call my mistress. Prefer bbw, black/white. Take it off, take it all off. Just ask and I will; like the true exhibitionist i am.

The following chat is discussing group get-togethers. In the submissive/dominant world get-togethers are a meeting of like-minded persons sharing a social night of their preferred lifestyle. There are actual clubs in cities that make their living supplying everything needed for one of these get-togethers, from dungeons, chains, floggers, to masks, swings, or clothing. I have discussed these clubs further in the Dominant Female section.

A CHATROOM CHAT:

RoseBud:	this was a D/s group you went to Wants???
WantsToBDom:	Yes, Rose... I think it was some new or visiting folks... or some who just didnt' give a rat's rear what happened to anyone
RoseBud:	geez... our group is very safety conscious
WantsToBDom:	OUr group is very safety conscious in it's rules... but some members are really into the extreme and some folks are new and don't know any better
RoseBud:	ignorance is a choice, not a condition
Love To:	I agree Mistress Rose
WantsToBDom:	Ture, Rose... and some folks seem to choose it more often with what they do at these meetings
RoseBud:	agreed
WantsToBDom:	Ithink our group has also had a recent influx of people loking for a quick lay
RoseBud:	arg... so common here too...they think its all about the sex
WantsToBDom:	I haven't beent o a meeting for some time, so maybe thigns have changed... I hope so...
Choose Me:	whatever Rose says as i gaze at her pic
Love To:	that is something I hear can happen...
RoseBud:	I gotta take that pic offa there LOL
WantsToBDom:	but there was also a very serious trend last year towards the fettish stuff...the kind you can get hurt with...not good
Mother4U:	Rose...i think you have an admirer!
RoseBud:	::hides behind Mother:::
Mother4U:	lol
Choose Me:	Ah just a short crush i'll get over it Mistress ☺
RoseBud:	lol
Love To:	I'm new to this too...I have been told there are a lot of wannabes and wanting quick lays, too
WantsToBDom:	Definitely somethign to watch out for, Love To, along with predators and bullys. Those that really get a kick out of inflicting the pain...they can be unrelenting in their punishment...makes for a bad morning after
Mother4U:	those kind of people are everywhere, no matter what you're into lol

Love To: very true Mother4U
Choose Me::   yup... and a quick lay isn't that bad!!!!
Love To: depends on situation and what you are looking
                                       for...but I prefer slow, steady, long.

NAME: MakeMe, **LOCATION:** PA, **STATS:** married male, **DESCRIPTION:** Looking to be slave for the day. Looking for mistress to use me. Forced exposure. Degredation. Dressing. brn shws gld shwrs real and phone submission humiliation wearing boots and speedo, farm fun

NAME: SubbieBoy, **CITY/STATE:** Chicago, IL, **STATS:** Single, 47 yr. old, **DESCRIPTION:** let me know if interested. Female adoration and submissive through a lifestyle not hobby. Houseboy in my "free" time. I am intereseted in developeing a long term relationship with a Dominant Mistress and let her lead me through life as she will be the one that knows what is best for her slave.

NAME: OnlyToServeU, **CITY/STATE:** NYC Area, **STATS:** 30 something, single, **DESCRIPTION:** call me Richard—or slave if that better suits you, Mistress. Real time submissive male on my knees ready to serve: to be bound, to be spanked and flogged, to worship your feet and body, to be forced feminized ready to be of service to you, Mistress. Mistress, how may I please you today??

NAME: Dolltoy, **CITY/STATE:** Bay area, CA, **STATS:** 5'8", 145, blue eyes, 33 yr old, **DESCRIPTION:** want to serve female, ts, couple or group very kinky. Love to lick and suck and beeing used for all your pleasures, like to watch and beeing displayed, straight executive in real lif. I am a nice person, very open minded, fit, uncut, partly shaved and sensitive, want to meet real people, pls. send me mail Use me as your sextoy for whatever you want, but please USE me. Sub, bi, bdsm, oral, femdom, humiliation, shemale, sissy, gs, mistress, dom, domination, master, tv, licking, facesitting, lingerie, showers no single men please

The following chat demonstrates how helpful chatroom regulars can be in helping to locate someone for a night of Internet sex, and also a humorous exchange between regulars when it comes to discussing/deciding on what a submissive should wear to please the dominatrix. As you may note also, it is not uncommon for a submissive to share his standing with the dominatrix with another submissive when they enter the chatroom. I myself, as a woman, have found this entire lifestyle extremely fascinating, and quite humorous, even though I know these people take it very seriously. There is just something about visualizing a slightly overweight, hairy, man dressed in a ladies corset and pink panties that tends to bring a smile to my lips and a small laugh to my throat.

A CHATROOM CHAT:

ComeToMe:	Lovely?
Lovely:	yes Mistress?
ComeToMe:	thought I lost you
ComeToMe:	:(
ComeToMe:	a scary thought—lay at my feet
Quickly:	any female looking for a male slave online tonight?
ComeToMe:	<YAWN>
Toallofu:	there MUST be Quickly...
Quickly:	may i be of service?
ComeToMe:	lol Toall
Toallofu:	no...i'm serious
Toallofu:	i just don't know who
Quickly:	anyone?
ComeToMe:	encouraging news for him I am sure
ComeToMe:	lol
Toallofu:	Denise?? you here?
SexyDenise:	hehe right now im very busy training :)
LovetobeAbused:	Chains
Quickly:	no one interested?
ComeToMe:	oh,...sorry Denise
CutiePie:	toys, toys, toys
ForgetmeNot:	Ms. Cutie, Ma'am, is this your color of pink?
SexyDenise:	oh geeze Forget let me get my sunglasses on
CutiePie:	not a good color, pet
ComeToMe:	<STREEEEEEEEETCHING>
CutiePie:	i want you to use SLUT pink tonight Forget. i want you as my slut female...
ForgetmeNot:	Ms. Cutie, I want to be as much like you as possible!
ComeToMe:	LOL Cutie
LovetobeAbused:	i'm in chainsssssssssssssssss
LovetobeAbused:	...oooOOOO
LovetobeAbused:	the echo off these dungeon walls
CutiePie:	think thats funny, Come?
LovetobeAbused:	i'm in chaiiiinnnnnnnnnsssssssssssss
ComeToMe:	bold is more like it Cutie

ForgetmeNot:	How is this Mistress Cutie?
CutiePie:	smiling...wonderful, my pet...you will be rewarded <gbg>
Steven:	Good evening, enchanting Mistress Cutie
CutiePie:	how do you like that, Come?
CutiePie:	hello Steven, my sweetie
LovetobeAbused:	i'm in chaaaaaaaaiiiiiiiiiiiiiinnnnnnnnnnnnnnns
Steven:	mmmmmm Hello, Ms. Cutie, my pleasure
ComeToMe:	I am not judging Cutie,...I just thought it was bold is all

NAME: Slaveslut, **CITY/STATE:** California, **STATS:** 30ish, single, uncollared, 6', 200lbs, **DESCRIPTION:** masculine submissive—slave, Searching for a Mistress to serve and please in every aspect. I'm a very good houseboy as well, cook and clean. Love bondage, collars, leashes, and I have a cute muscle butt for spankings, paddling, str*pon d*ldo's and an*l toys. I love to massage a lady as well. On my knees, Mistress your wish is my command.

NAME: Belong to My Mistress, **CITY/STATE:** Indianapolis, IN, **STATS:** Male, birthdays annually, **DESCRIPTION:** Fullfilling the needs, wants, desires, and dreams of the one I love and serve, Mistress Barbara. Finding the line between Pain and Pleasure and taking it one step further with firm/loving hands 0/our minds...Is that not the place where all fantasy and reality being Mistress?? I am a slave... AVAILABLE FOR USE BY OTHERS...contact Mistress Barbara. People who have found happiness evaluate themselves. People who are unhappy evaluate and criticize others lives. So be happy, be free, be real.

NAME: SUBBIETOSERVE, **CITY/STATE:** louisville, KY, **STATS:** male uncollared, **DESCRIPTION:** looking for serious D/s relationship. I want to serve a Mistress R/T in my area who has a desire to train me to serve her in anyway she desires. I am a total submissive in every sense of the word and desire only to please. Interests include but not limited to: B&D, S&M, CBT, nipple clamps, Humiliation, oral and an*l play. Mail me if interested. Serious replies only please. Collar me??

NAME: ShoeLover, **CITY/STATE:** Peterborough, England, **STATS:** Male, 42, **DESCRIPTION:** Looking for a mistress who understands boot and shoe fetish, Boot licking, enjoy being trampled and kicked (now can you rub those sore bits?) and generally being a sub to a beautiful female. I make enough money to buy gifts (boots, shoes, &sandals) for the right woman. (Where is she?) Is it too much to ask?...to find a female who can love and care and... understand sexual fetish and submissive needs?

NAME: whatever I'm ordered to answer to, **CITY/STATE:** Chicago area, **STATS:** 6'158, 38b-29-38, **DESCRIPTION:** love being on a short leash at my Mistress's feet... being a dominated submissive sissy. Submissive TV sissy, long legs, sensual, sweet, but at times can be wickedly naughty... Like being placed in a tightly laced black satin Victorian corset, sheer black hose, 5" heels, a satin pinafore, and a black leather collar. Sissy Maid- domestic duties, chambermaid, massuese, nanny, pedicures, manicures, laundry wench, sissy sl*t service... Adore corset training, Ds, BDSM, and soft silky pretty things. Totally subservient sissygirl to my beautiful Mistress. I am her possession, her maid, and her adoring exotic plaything.

NAME: SUBBOYTOY, **CITY/STATE:** Willow Grove, PA, **STATS:** Male, 21, **DESCRIPTION:** Looking for Sincere Dom. BDSM, Bondage, Domination, Wearing Sexy lingere, being treated like the sub-bitch that I am. Tied me to anything & everything. Gag and blindfold me. Force to me perform orally. Pull around by a leash attached to my collar. Push me around and degrade m I love to wear skin tight Latex, leather, rubber & then abused. Gently spank & whip my ass & chest & pull & sqeeze my nipples. Submissive boy-slave. Mail me if interested! I like to do things spur of the moment! Please Cum all over me! "Yes Mistress," "MMMPH" <- I'm gagged "Cum in my mouth!" I love to suck guys off even though I am not gay. So if you are close and want a BJ, Just mail me.

In following links to web sites on submissive lifestyles I was perplexed as to why this lifestyle is dominated mainly by upper professional men. This being the case, it tends to make me believe that after a hard day at the office some executives really do not want to make one more damn decision, and prefer just being told what to do by someone else. Could this be the cause for the high rate of professionals in this lifestyle?

NAME: YourPleasure, **CITY/STATE:** Columbia, SC, **STATS:** uncollared and single, 31, white; 6'; 190 lbs; brown hair; hazel eyes; clean shaven; smooth; great smile with dimples!, **DESCRIPTION:** personal humiliation through: D/s, human "puppy" and "horse"/"pony" training, domestic/maid/houseboy(girl ;-)) service; desire training, discipline, spankings, forced cross dressing/nudity and chastity; diapers and age play; and many more interests. "Please _____(Mistress), may I?"

NAME: loveoral, **CITY/STATE:** Florida, **STATS:** 38yr ol, male, single, **DESCRIPTION:** submissive seeking openminded, uninhibited, sexy female.

Love kink, love GS's and oral body worship. Mistress I am yours if you want me. Let my face be your saddle. Open to training.

Below is a submissive who wants to return to diapers and being a 2 year old. I was absolutely amazed and fascinated, during my travels, at how many men want to revert to their infancy. I had witnessed this lifestyle once on a television show, but thought the whole thing was staged. It is apparent now that it was not. I don't believe I know many women who would even consider this type of lifestyle for more than 30 seconds, let alone a lifetime.

NAME: Wants2Bbaby, **CITY/STATE:** Chicago, **STATS:** Male, 42yrs young, 5'9 150 blnd/blue 30"waist, divorced, **DESCRIPTION:** being submissive and being a baby again...seeking to forced into diapers fulltime again... have to change myself ☺ put me in a diaper and regress me back in time to the terrable 2's and my life can start anew

NAME: UseMeAsUPlease, **CITY/STATE:** can relocate, **STATS:** male, **DESCRIPTION:** extremely submissive male in search of a Mistress to serve. Very much into oral servitude, lots of anal play (strap-ons & vibrators), humiliation, restraints, golden showers & enemas, spankings & paddling. Would consider groups, for the right Domme. Poss cpl

Here is a chat explaining a punishment level for the submissive, in a subtle way. Most dominant women tend to speak in innuendoes with their submissive, but even so, the point is well taken.

A CHATROOM CHAT:

MistressDi:	I don't require chastity of my subs, it's just a lot healthier for you if they are :)
MerryMistress:	true, but i push them to that level, so they know ahead of time what I expect. Let the punishment fit so they understand.
Sid442:	why so, MistressDi?
MerryMistress:	oops :)
Slave2Day:	Just because you're a Domme doesnt mean well, I have my own opinions to which Im entitled as a "human being" first ;o)
MistressDi:	It's not a smart move to piss off a woman who carries a whip and tends to run low on patience AND estrogen :)

MistressDi: Slave2 I totally agree with you, but I still require a high level of respect from my subs! Without the respect one loses the control and its all about control;-)

NAME: YourSissy, **CITY/STATE:** Tampa Bay, St. Petersburg, clearwater, FL, **STATS:** Male, Divorced, 6' 175lbs in good shape, 40, **DESCRIPTION:** desire training, discipline, spankings, forced cross dressing/nudity force bi encounters to please my mistress and chastity; submission, recreational humiliation, roleplaying fantasies and many more interests. All women are welcome and I respond to all women. Would like to meet a true attractive pre op shemale. On my knees before Her with head bowed, I felt Her reach behind me as She fastened the collar around my neck...and I was Hers.

NAME: NJSubClaimed, **CITY/STATE:** New Jersey, **STATS:** Male, Claimed, **DESCRIPTION:** Worshipping Mistress Erica's heart, mind, body, soul and feet, pleasing Her, being romatically submissive with the Goddess of my dreams. Showing my Mistress that She is the most beautiful Woman in this world! Serving Her! Being a good lil boy for Her! Life will be so wonderful with Loving Toe Goddess to guide, lead, discipline, cherish and nurture me! I am lost without Her. The hunger I feel for serving my superior queen leaves no doubt that my place is at Her feet. Withouth Her, I am empty. There is a magic in the melding of a loving Mistress and her loyal needy, loving pet!

NAME: ForeverAtYourFeet, **City/State:** Phoenix, Mesa, Tempe, Chandler, Scottsdale Arizona AZ, **STATS:** Mid 30s, Male, single, available, **DESCRIPTION:** Total submission to beautiful woman. BDSM, Bondage. Looking for an attractive RT Dominatrix/ Mistress / Domme for pain, pleasure, and service. spAnk, slap, whip, suspension, tease, torment, torture, kidnap, handcuff. High pain tolerance experianced. Houseboy, sex slave. "Tie me up & put me on my knees...and keep me there."

NAME: Palace Slave, **CITY/STATE:** New Orleans, LA, **STATS:** Single, **DESCRIPTION:** Not a hobbie but desires and destiny...To be an oral Puppy slave to any Mistress who will have me. Put a collard around my neck and lead me around to service you orally any where, time, place and as long as you command me to perform orally to anyone you order me to, Be it your feet, or whatever, use my face as your personal seat I am yours to command. I feel like I am the only and lonliest black male submissive in the world all I seek is a Mistress who would let me sevre her

NAME: SlaveAdventure, **CITY/STATE:** Los Angeles, CA, **STATS:** Male, committed, **DESCRIPTION:** Totally open & unpossessive Submission, SM, bondage, obedience, service, erotic pain…and just about anything else you might desire, Ma'am! Looking for an intelligent, personable, experienced, demanding Mistress with a sense of humor. Can serve as single, evenings and some weekends, and / or…bifem committed partner can join as You may please. Nothing is kept a secret between us.

Excuse me, but being a woman I just need to say something about the following profile. "Is this guy kidding?" It's quite obvious he has never had a "dinner" date wearing a tight corset, girdle, or TIGHT pants! You want to discuss REAL torture?

NAME: tighterisbetter, **CITY/STATE:** NYC, **STATS:** male, **DESCRIPTION:** submissive, meeting other cd"s. I am a crossdresser who loves tight clothes, girdles, corsets, heels, fetish fashions. Very passable. Can be Domm too. Fantasy of forced abduction, kidnapping, as a woman. A forced wedding sceneerio. The TIGHTER, the BETTER. Not so tight! Mummmph! Looking for a tv/cd Mistress to photograph me in various bondage positions. Ballgags, trainers, straitjackets, etc.

NAME: PlayMeHard, **CITY/STATE:** Seattle, Everett, WA, **STATS:** Male, excellant shape, 6'2" 215# very clean cut, 30's, **DESCRIPTION:** BDSM from mild to edge play…Dom, Master, Domme, Mistress, submissive, sub, slave, sadist, maso, bonda#ge, electrotorture, roleplay, dungeon, whip, torture, interrogation…and? I am a maso/slave/sub iso a Sadist Mistress for RT play! From mild to edge play, it's all good! Need a boy to play with??

Following is a chatroom chat that took place on a Friday night. I accidentally stumbled upon this chatroom which had submissives "servicing" doms. Needless to say, I didn't say a word, just sat back, put my feet up, ordered a pizza, had a beer, and watched…quite an interesting Friday night on the computer screen.

A CHATROOM CHAT:

LadyBeGood:	nipple taunt... hard
ToServeU:	biting it
ToServeU:	licking it
ToServeU:	squeezing it in my hands
ToServeU:	moaning as seeing your pleasure
LadyBeGood:	yessssssssssssssss, ToServe

LadyBeGood:	yessssssssssssssss
LadyBeGood:	maoning softly...
ToServeU:	loving the whole breast
ToServeU:	massaging
ToServeU:	looking up and kissing her hard again
LadyBeGood:	lips firmly planted on yurs...
ToServeU:	<grinning and pulling away>
LadyBeGood:	growling... this is how you serve your mistress?
ToServeU:	moving down to you other breast
LadyBeGood:	a low deep growl...
ToServeU:	biting it hard as she growls
LadyBeGood:	oooooooooooohhhhhhhhh
ToServeU:	gripping it with both hands and squeezing
ToServeU:	wetting your nipple eagerly
LadyBeGood:	soothing my growl to a purr...you please me ;-)
ToServeU:	sucking you hard and tight
ToServeU:	my tongue squeezing it
LadyBeGood:	grinding my ass down on yur legs...
ToServeU:	burying my face in your breast
LadyBeGood:	my moistness against you...
ToServeU:	swallowing you whole
LadyBeGood:	ogoddd
ToServeU:	biting your nipple again
ToServeU:	circling it with my tongue
LadyBeGood:	chills up and down my spine...
LadyBeGood:	tingling...
ToServeU:	licking Ladys feet <grin>
MistressToy:	Lady - if i walked up behind you while Toserve was suckingon your breast, massaged your neck...
MistressToy:	would you mind?
MistressToy:	if I then kissed your neck softly using the tip of my tongue
LadyBeGood:	mmmmmmmmmmmmmmmm <grining> to feel my whip is true delight, Toy
ToServeU:	<grinning at Lady>
Contrl4U:	can I get my pussy licked by a sub...all this hot sex talk is making me wet
MistressToy:	i willdo you Mistress Contrl...
Contrl4U:	do me Toy...great
ToServeU:	<grinning and sucking LadyBGood into my mouth again>

34

MistressToy:	licky you Contrl...lol
LadyBeGood:	<giggles>
ToServeU:	<squeezing with BOTH hands>
LadyBeGood:	omgggggggg
MistressToy:	sitting down
Contrl4U:	yes, with my legs wide open...touching my wet pussy—get moving on me Toy!!
ToServeU:	mmmmmmm as she looks at me and whip again.
MistressToy:	your tits are aching for me Mistress Contrl
MistressToy:	as is your wet pussy
ToServeU:	<running my hands up to your neck>
MistressToy:	i know you will train me tongight
LadyBeGood:	eyes locking onto yurs...
ToServeU:	pulling your face down to me
Contrl4U:	oh, but of course I will <g>
ToServeU:	licking your lips
LadyBeGood:	mmmmmmmmmmmmmmmm smiling wickedly.
ToServeU:	sucking you into my mouth
Contrl4U:	touch me
MistressToy:	you want a tongue...and i certainly have one
Contrl4U:	it's really wet
MistressToy:	so i feel
ToServeU:	exploring you fervently
Contrl4U:	on my clit—kneel you slave
MistressToy:	I kneel before you
LadyBeGood:	the kissrowing passionate...
MistressToy:	licking oyur knees
Contrl4U:	I spread for u
LadyBeGood:	kiss growing*
MistressToy:	working my way slowly down
ToServeU:	<soft low, deep kisses
Contrl4U:	mmm
MistressToy:	towards your pussy
Contrl4U:	mmmmmm
LadyBeGood:	ohyessss
MistressToy:	i reach up to twist a nipple
Contrl4U:	gently! Easy...
ToServeU:	biting your lip and pulling it out
MistressToy:	hear you groan
ToServeU:	grinning at you
MistressToy:	you grab myhead
ToServeU:	you are SO fucking hot, my lady

Contrl4U:	ooooooooooohhhhhhhh
MistressToy:	wanting to puch my face into your pussy
Contrl4U:	grabbing
Contrl4U:	ohh you think you are gonna get away with this huh?
Contrl4U:	\<smirking\>
MistressToy:	but i want you to wait
ToServeU:	\<grinning innocently\> huh?
MistressToy:	to tease you
Contrl4U:	no...you WILL do it now!!!
Contrl4U:	sliding my hand down between us...
MistressToy:	i feel your hand move down to my cock
Contrl4U:	finding you...
Contrl4U:	i want to twist you…to reshape u
MistressToy:	\<gasping\>
Contrl4U:	this is Mine...to use as I please \<g\>
Contrl4U:	i am going to put a ring on your cock so you won't forget who is in charge here
MistressToy:	yes, my mistress, I understand
LadyBeGood:	rubbing along the shaft of ToServe...
ToServeU:	soft moans
LadyBeGood:	attaching clamp to ToServe dick
ToServeU:	whimpering softly
LadyBeGood:	moving your dick between us...
MistressToy:	rubbing your cli\t my lady
Contrl4U:	u need to fuck me...NOW
ToServeU:	holding you down by your hips
Contrl4U:	in & out and lick me NOW

Our journey has now ended in the world of the submissive straight male. Submissive men come in all shapes and sizes, from varying walks of life, and could be any man we know. They believe and live this lifestyle to its fullest, and really do worship the dominant women in their lives. I know this doesn't sound very masculine, but I'm sure some would find it extremely masculine that a man could adore a woman in such a way. Most women I know have never experienced being adored on a daily basis. It is only something they have read or dreamt about, and yet it does seem to exist if only in a slightly different manner.

STR8 DOMINATORS

Here lies the other side of the coin in our submissive/dominator lifestyle for straight men. In the previous chapter we discovered the straight male submissive role of being dominated by a female, but here we will glimpse into the lives of the men who are the dominators of women. Dominators of women who love, worship, and have willingly relinquished all power to their dominator.

The dominator expects full obedience from his submissive and will gladly inflict punishment if that obedience is not met. The submissive female in this lifestyle must understand that the punishment is a sign of the dominator's love and willingly accept the punishment, thus proving her total trust and loyalty to the dominator. The dominator will make all choices for the submissive's life, and once again, this is a chosen lifestyle, one in which both partners have freely agreed to experience together.

Most of the male dominators in these chatrooms will openly chat with you about their lifestyle, but usually only in a private chatroom. I chose not to use any conversations from private chatrooms, as I felt these conversations were not for public display as they were given to me in trust. I did find most of these men well educated and quite interesting to speak with, even though I do not agree with this lifestyle and it's brutality.

Most of these men take this lifestyle very serious and are extremely adept at mind control through the use of the English language. Words seem to be spoken at the just the right time in order to obtain the mind-set from the submissive that the dominator desires. This is true in real time as well as on the Internet. If you note, the majority of profiles written by dominating men are written in a style of soft, romantic, gossamer-wing sentences. The men also portray enjoying nothing but the finest that life has to offer—at least in their profiles, i.e., classical music, fine wine and fine dining. The picture they paint in ones mind when reading the words, is of a wealthy land baron, living on an estate with nothing but the finest surrounding him, or a knight from medieval times looking to rescue his lady who is locked in the castle tower. Who could not fall for such romantic words written in such a way as to take one to another life, another time. Oh, the power of the written word!

Let us mount our trusted steed and venture forth into another time, another place, slaying dragons on our way, and explore the profiles and chats of the dominating man as he searches for his lady in the tower.

Donna Tracy

NAME: Sir David, **CITY/STATE:** MA, RI, NJ, NY, NH, VT, CT, MD, ME, DC, **STATS:** 45, Divorced, fit, handsome, **DESCRIPTION:** Dominant trainer and disciplinarian of intelligent, attractive, (slender, leggy, petite, or proportionate), sexual, submissive women 19-45 sgl, div, married, sirens, slaves, and angels. Noviciates to the hardened "Edge". Realtime only. I enjoy the submissive females in my service, the arts, classical music, literature, photography, painting. Lifestyle "Master" of submissive females, bixesual females, lesbians, and switch/domme females S&M, D&S, B&D are at the fascinating edge of human sexuality. "Not interested in your gift of submission, will accept only your total surrender "The Story Of "O" exists"

NAME: Sir Stewart, **CITY/STATE:** Winter Haven, FL, **STATS:** Male, **DESCRIPTION:** Come to me my little one. Yhur pleasures and desires are longing to be awakened. None is needed, as she is connected to me and knows my every thought. Come to me my little one. You will always have my passion, compassion and protection always. She was lost, I lit the way—She was unsure and I gave her conviction—she reached out her hand and I took all of her.

NAME: Lord Seeker, **CITY/STATE:** Your Dreams, **DESCRIPTION:** I am He that you seek... Your Ultimate Sin... Your Darkest Dream... The Deep Driving Force in the midst of your Scream... I own your Soul. I caress your Fear. I am He, The Master your crave... and you... little one... are my sweet darling slave. 'fore the hearth... on black fur you play in the light of the flame... 'til summoned by Me, and the sound of the chain... and come on your knees... 'cross the stone to my chair... and savor with lust what awaits you there. And then, in my arms, to the Altar... the thrill of cold marble... the caress of the halter. And blinded and bound and spread out before Me... your quivering flesh will beg for the leather... that may sting like a bee, or kiss like a feather... awash in your heat, now pulsing, now pounding, erupting within, you shall face and embrace your Ultimate Sin. Now writhing and lost in the throes of your passion... you shall have all you crave... for I am your Master and you... little one... are my sweet darling slave—

NAME: MasterGift, **CITY/STATE:** NYC, **STATS:** Male, 38, **DESCRIPTION:** 17 years experience in BDSM, leather, chain, ropes, wood, canes, simple things very complicated feelings. Looking for a very sub "Female", to teach discipline with an accent, self control, looking for R/T!! Please me and I will take very good care of you... Riding crop and toys... Master...Dominant...Singing Opera...a glass of red wine...Couples/Dom's/Domme's & sub's looking for advice wellcome!!!! be a fem, or a bifem, be submissive, very submissive, then mail me, NO MEN!! Phone/Voice verification a Must!!! no pic or profile no answer. Open your mind and trust me, then you will be mine....and I will be pleased. "Tomorrow is promise to no one"...

In case you have started to wonder, I really did not write any of these profiles. They are all originals by their creator, whoever he/she may be. I will let you in on a little secret—after following some of these men's web links, I couldn't help but sit here and have a good laugh with myself. It's not that I have a warped sense of humor, well maybe just a little, but here we have writings of wanting, lust, desire, and control by dominant men. But in checking out their photos on their web pages, somehow their photo didn't quite fit the image they are portraying in their profile. Now, don't get me wrong, I have absolutely nothing against a slightly pot-bellied male, but when you see him standing there in leather, waving a riding crop, with a slightly receding hairline—well, it's all I can do to contain my laughter. After all, from the words they had written I was expecting a Clark Gable or Errol Flynn to be the one rescuing me. Somehow I just can't take this very seriously when there is an actual photo attached to the profile. But remember, to the authors who wrote these profiles, it is very serious.

NAME: Trustofmind, **CITY/STATE:** of mind, **STATS:** Male, 31 yrs, **DESCRIPTION:** she looked out the window as the rain fell to the sill, the drops splashing and falling to her lips, which reminded her that the raindrops tasted like her tears but without the pain. ...with her arms tied to the poles at her side, the tip of his wing traced her neck softly... a gasp escapes her lips and her eyes close on their own... he quickly took it away, teasing her. He circled her restrained body secretly ravaging every inch of her with his eyes, leaning over her shoulder from behind; he growls softly into her ear, "Tell me? Why I should touch you again?" "Remember, if you can... the way you moved when our hands touched, I remember the taste of a love so pure it was deceiving" "Good Kitten"

NAME: Crop Master, **CITY/STATE:** MD / DC / VA, **STATS:** Male, **DESCRIPTION:** Experienced r/t Master, rake, rakehell, roue, dominant. Molding deliciously intelligent women into moaning instruments of pleasure. When your body starts aching for the whip, it's really your mind that is longing for release. Theatre, film, literature, art, aesthetics. Laughter, ecstasy, refined pleasures with a touch of barbarism. Tempt me.

I am finding this play on words quite interesting and am beginning to question why most of these men seem to be living in the major cities of the country. Could it be large cities are a more "civilized" area for such a lifestyle? I do know the clubs that cater to the D/S lifestyle are based mainly in major metropolitan areas—at least I don't think I have heard of one in Lubbock, Texas, but once again I could be wrong.

NAME: ControlwithCare, **CITY/STATE:** NYC MetroArea, **STATS:** Male, born not yesterday ☺, **DESCRIPTION:** RealTime D&S/B&D (V)aster. The RealTime world of D&S/B&D ONLY! NO cyber "play" no mails begging for it! If u are cyber I respect that but not for me! Mail me at XXXXXX 1st in subject line; 1 face, 1 Full Length (non-sex) pics: BE responsive to this profile. Master & U can ONLY Serve 1. LegalYoung-YoungishMid slim/attractive fems knowing: their inner little girl; her need to serve, mutually grow; submission is given not taken; strong, intelligent women make the best slaves. NO "bedroom subs" OR NeverWillBees. WANT2Bees OK. For 1st contact/maintenance Spare me cyber "lists"/"rules", NOT a cyber "dom" I have a clue of responsibility to learn u D&S, B&D usually within the context of a D&S relationship. Hierarchal D&S Reality: Fems Domme Fems in Mine. Cruel to be kind in just the right measure…Hands hard as steel for training/discipline, soft as silk when deserved…Your hands are small I know, but they're not mine they are your own and you will not be broken, "cause in the end only kindness matters".

NAME: Master Will Do, **CITY/STATE:** New York City area, **STATS:** Male, **DESCRIPTION:** WOMEN ONLY!!!!!!!! IF I DON'T ANWSER I'M NOT HERE…PLEASE MAIL ME…I do want to talk to you!!!! looking for a good R/T sub…Loving D/S, romance with it, self-growth, day-trips, photography/video, my big country home, quiet time too, dancin, going out for fun, you??!!…and for those that ask: yes I do allow young/legal FEMALE sub, Attitude & Willingness much more important… I am a Therapist, mental health…and yes I do clinical sex therapy for real…Clinical/Medical Hypnotherapist. D/S is for expanding love and intimacy and trust…not a Fetish category.

A CHATROOM CHAT:

Belongs to Sir:	explain to LadyImperial about what the Sjambok looks like Sir
LadyImperial:	i think i got the picture Belongs
Belongs to Sir:	lol ok
LadyImperial:	it is what it feels like that is more an issue!
Sir Blacktail:	lol, saves me trying to think so early in the day
Belongs to Sir:	figured maybe DeeDee could get a different perspective since DeeDee watched me get whacked with it
LadyImperial:	She did?
Sir Blacktail:	oh I will never forget it
Belongs to Sir:	she saw one hit

40

Sir Blacktail:	one was bad enough
Belongs to Sir:	he gave me another after she left
Belongs to Sir:	for good measure
Belongs to Sir:	she also got to beat my ass, lol
Sir Blacktail:	lol
LadyImperial:	did he do anything nice afterwards?
Sir Blacktail:	<—has to learn to keep eyes open when beating someone
Sir Blacktail:	lol
Just for you:	whats nicer than a whack Belongs
Belongs to Sir:	sure did Lady :-) we had a good night
LadyImperial:	Just! lots of things!
Sir Blacktail:	yes it was
Sir Blacktail:	I enjoyed it :o)
Just for you:	lolol
Belongs to Sir:	i'm glad you did :-)
Belongs to Sir:	even though you felt bad for beating me, lol
Sir Blacktail:	still do
Sir Blacktail:	lol

NAME: Master Richard, **CITY/STATE:** Georgia, **STATS:** 44 Years Old, Male, Seeking, **DESCRIPTION:** What is pleasure without pain? What is intimacy without kink? What is the point of living if one is not truly alive? Slender and pretty, she awaits her Master, the one who will own her totally, free her, use her, give her joy, happiness and contentment. She sits and looks off into space feeling that the empty spot without her Master is larger than all other spaces in her combined. Go fetch her and tell her her Master has arrived.

NAME: MasterSeekingU, **CITY/STATE:** VA, MD, **STATS:** Male, single, 54, **DESCRIPTION:** Taking a submissive to her limits of sensual endurance. A tall dark handsome professional Master. Seeking the sublime... the frolic nature of a self respecting submissive. The path to R/T may be unusual but in the end it should be rewarding. Hoping to create our own realm. Yes I keep a crop laying on my desk for you to retrieve. Submission is a gift from the heart. To subscribe to others labels is a venture into unhappiness. Be yourself. OTK. When my hand cups the underside of your breast the control begins. When my hand stings the cheeks of your bottom the discipline starts.

It is my hope that parents who are reading this book might see how simple it could be for your young daughters to be brought into this lifestyle. Read the words that these men use and imagine a young, naive, teenage girl who believes no one cares about her, finding some of these profiles on the Internet. Here is a man who has nothing but her interest at heart, who cares deeply for her, would make her his one and only, and shower her with mounds of love. The words are the bait - the rest is the training.

NAME: BeGood, **CITY/STATE:** Florida, **STATS:** Male, Single, **DESCRIPTION:** Looking for that little girl who needs her daddy's guidance and needs his firm hand applied to her bare bottom when she misbehaves. She tries to be good but can't seem to quite master the concept. Needs to me scolded, lines, corner and above all else is by nature a little girl. (be over age 18 young ladies) Understand that daddy just wants her to be the best little girl she can be. Young lady you are now accountable for your actions, go fetch the brush and belt. "yes, daddy"

NAME: WifeIsSlave, **CITY/STATE:** Philadelphia, PA, **STATS:** Male, 31, **DESCRIPTION:** BDSM, D/s, B&D, S&M, Dom/sub. Not a "lifestyle" dominant, just really enjoy the play. I specialize in wax play, but enjoy all forms of inflicting pain/pleasure. Have a fantasy you want to explore with a normal guy? Mail me. Use Mail to contact me. NO PROFILE = NO CHAT, NO EXCEPTIONS. NO MEN EVER! Trainer of submissive females. R/T only. My wife is my best slave. Are you a Dom/Domme? Inquire about my wife. I know it hurts, but you like it that way.

NAME: Dom to treat, **CITY/STATE:** around, **STATS:** Male, single, Born in July, **DESCRIPTION:** I believe very deeply in True Lover and it is something that I think is worth fighting for, but not dying for. Anyone can die for any reason...I'd much rather live for love. Of cours, so few people believe in love, these days, and it is truly a shame. If only more could understand the beauty, simplicity, complexity, naivety, and wisdom of love, this world would truly be a glorious place in which to live and share. Indeed, romance would flourish and passion would abount...but, alas, the world is filled with hate, greed, ignorance, intolerance, selfishness, and those who use others for nothing more than entertainment. Tis truly a shameful world in which we exist, but I shall continue striving to truly Live and attain True Love. For this is the most noble of quests...though, it is long and painful, it will be met with the most wonderful treasure and none in the world shall know such peace as I.

NAME: Master Trainer, **CITY/STATE:** Los Angeles, CA, **STATS:** 37 yrs old, 6', 190 lbs. 43 inch chest 32 inch waist, **DESCRIPTION:** STRICT but fair & caring Dom. Very submissive females who are...intelligent... attractive and who are into b/d, d/s, s/m, and other initials. I also practice Martial Arts. Tormentor...and teasor...spanker...and whipper...humilliator "Daddy promises...this will only hurt...a little".

A CHATROOM CHAT:

ForUSir:	<~~loves to be TOLD to swallow
ForUSir:	lol
IObey:	:::dancing for Squire:::
Squire Manor:	ForU, swallow!
ForUSir:	mmmmmmm
NoLady:	ForUSir????
ForUSir:	hmm speaking of that I'm thirsty
ForUSir:	yes No Lady?
NoLady:	swallow and shut the f*ck up!!!!
ForUSir:	LMAO
NoLady:	:)
Master of ALL:	<— reminds Myself to tell the girl to swallow the Dom Perignon 76... on command
Lord Easy:	...one would think that if one was swallowing, talking would be impossible anyway...
Squire Manor:	wiggle it Obey
IObey:	::::wiggling for Squire:::
Squire Manor:	not in ventriloquism submission Lord
Lord Easy:	ROFL
IObey:	:::swaying close to Squire:::
Lord Easy:	blowjob by proxy?
IObey:	lol Lord Easy
IObey:	thank You Sir
WantingU:	o no that wont do
Squire Manor:	put her on your knee gag her and make her say whatever you want
IObey:	i'm in chains
Squire Manor:	lol
Lord Easy:	i LIKE that, Squire... all the fun and none of the fuss

NAME: SubTrainer, **CITY/STATE:** Long Beach, CA, **STATS:** Male, **DESCRIPTION:** Exquisite training in submission for the willing female. Immersion or casual; first the mind, then unrestrained devotion... and training. Ownership the goal. 12 years in the scene, a lifetime in the roles. Understanding, depth, and willing submission of your soul. Chatrooms are barely tolerable; I'll watch, but speak rarely. Mail is welcome from true willings. Accept, yield, respond, and beauty you will be.

NAME: KnightofOld, **CITY/STATE:** in your heart, **STATS:** Male, birthday long ago, **DESCRIPTION:** I reside within the heart and soul of my precious one, who has chosen to give freely her submission and her love to this content one ☺ I continue the study of my own dominance and only then may i be able to take the raw passion that is within the female and mold into the perfect blossom that she so desires and craves, within that blossom will she thrive and be forever content ☺ I live on a plateau created for me by my precious one and i am very happy with the servitude, love, and desire that abound ☺ only in this place of complete peace may i be able to warm her heart with my loving hands and my gentle caress for she has given to me the breath of life and the gift of the most treasured ::: submission…my respect always and my love to the precious one ☺

NAME: MasterStewart, **CITY/STATE:** Alabama, **STATS:** 45, male, Married to a vanilla spouse!!, **DESCRIPTION:** Looking for REL Time older woman married or single for real time D/S Discretion is given and expected. Lets explore your limits together as you submit to your desires. The sting from my hands is your gift! Enjoy it, I may give you more!!

NAME: Mentor4U, **CITY/STATE:** Your neighborhood, **STATS:** Male, born in July, Married, **DESCRIPTION:** Providing protection and guidance during explorations of those dark areas that you've yearned to investigate. An accomplished spanker, disciplinarian and mentor—a light-hearted style, but with a special interest in working and playing with intellectually exceptional women whose bedroom and boardroom lives reflect the breadth of their beings. Females only, please. Please feel free to mail me, but…………No profile, No chat…No kidding!!!!

NAME: SquireofSenses, **CITY/STATE:** South Jersey Shoreline, **STATS:** Late 30s, single, male, **DESCRIPTION:** esoteric pursuits of the mind…which control the flesh…transending power exchange…sensual dominance...body shaping ….mind expansion…endorphine rushes…mindmelding …riding the wave of the raptures of sublime dominance and submission idol maker...dreak maker ☺ erupting ones' limits setting the soul a blase with desire. Those who can't laugh at themselves leave the job to others…good enough ner is...

A fun evening in a dominant chatroom with teasing of a new male member to the chatroom, is printed below. While fun and games are played on the surface, I really wonder what the dominant male mind is planning on the inside for his next submissive?

A CHATROOM CHAT:

Misty Robin:	hiya Gentleman...Worth said we HAD to be nic to you
LadyNice&Easy:	Misty only you had to be nic
Gentleman2U:	or what, Misty?:-)
I'm Worth It:	i forewarned them you were my friend,
I'm Worth It:	Gentleman...and to be nice to you =)
Misty Robin:	nice my e sticks
Gentleman2U:	Thank you, WorthIt!!
TearDrop:	What was the term that was used by I'mWorthIt? uh... mmmmm "fresh meat"...
Misty Robin:	im not asking Gentleman
TearDrop:	Or was that my wording...
I'm Worth It:	ROFLMAO...NOT Tear
LadyNice&Easy:	::sending Misty a new e::
I'm Worth It:	that was YOUR term
Misty Robin:	thanks
Gentleman2U:	Fresh meat, ehhh?:-)
TakeMeSlow:	DomforLove do i have to give ya a grope
TakeMeSlow:	LOL
TearDrop:	Me?? Not me...
TakeMeSlow:	<~is rope
TearDrop:	I would never ever ever say anything like that... Dont even know what it means.
TearDrop:	=)
I'm Worth It:	that's some profile you have, Gentleman
DomforLove:	LOL i hope not LOL
Gentleman2U:	I'm a man of mystery, WorthIT, and of few words:-)
I'm Worth It:	i don't like it
I'm Worth It:	lol
Gentleman2U:	your comments have been noted:-)
I'm Worth It:	thanks
I'm Worth It:	<wondering now what Rope's sexual preferences are>
I'm Worth It:	men or women

Here we have the dominator who makes it clear that if you do not understand why he is entitled to more than one submissive in his life, then you have no idea what submission is. Right away he tries to make the submissive feel stupid and

45

incompetent, while trying to place himself in a much more knowing, higher position. Seems I have heard this same explanation used by bigamists.

NAME: Master of Limitations, **CITY/STATE:** Ft. Worth, TX, **STATS:** Male, 46 yr old, **DESCRIPTION:** If you cant serve a multi-sub DOM you do NOT understand submission. Vanilla Romance weakens D/s. If its NOT...SAFE, SANE, &CONSENSUAL...its NOT D/s. Im an RT Training DOM with 6 years RT experience...PLAYsubs do NOT need apply. Beware...Screening Tests used to help eliminate the playtypes. A True DOM knows what the "R" word means and lives by it always.

NAME: ComeToMeSoftly, **CITY/STATE:** Augusta, GA, **STATS:** Male, single, 46, **DESCRIPTION:** BDSM...to receive the power that is exchanged to Me and to turn it into the needs that she desires. To control dominate and lead her to the places she desires. The hand that spanks you sets you free.

The following two profiles caught my attention in that they both seem to go to the extreme for pain. The first profile clearly states what he is into, but the second profile is very subtle and could be the "spider to the fly" scenario. Both profiles show just how extreme this lifestyle can be even though it is lived in the name of "love".

NAME: MasterExtreme, **CITY/STATE:** New York State USA, **STATS:** Male, single, 40, **DESCRIPTION:** ALL THE TOYS FOR PAIN AND PLEASURE AND IF I AINT GOT IT I LL GET IT OR ITS NOT MADE Master of pain - enemas—hoods- gags an—restraints- clamps—cuffs featuring Enemas TT PT Wax Needles Whips humiliation degrading wax needles females only s&m Bd cages gags rubber and latex wear I do it all except blood scat kidsexams given 24 hrs aday by appointment mail me anytime.

NAME: IinflictThePain, **CITY/STATE:** Western Suburbs of Chicago, **STATS:** Male, Single, 49, **DESCRIPTION:** Real Time Play, Fantasy Role Play Sensual Interrogation. Mail welcome. I seek an Intelligent, Self-confident, Woman. With a Feminine look. Slim to Hour-glass figure To explore and try to"Dance on the Edge of a Razor". She will Trust that she will come to no harm while she spirals down in Desire, Panic, Passion, and Sensuality. I will Share with her Truth, Trust, courage, Respect, While she is reduced to a cherished object of Sexual bondage and torture compelling a release of her Darkest most secrect needs and desires "For how can man Die better, than facing fearful odds, for the ashes of his Father, and the Temple of his Gods.?"

And, here we have the "nice guy" just looking for someone to cuddle with and live happily-ever-after with. Somehow I have a hard time believing he is really into heavy domination, but then again I could be wrong. What do they say? "Let the buyer beware!"

NAME: SirDom, **CITY/STATE:** Boston, MA area, Southern NH, **STATS:** 36, White, 6'1", Male, Single ☹ **DESCRIPTION:** Hi, You can call me Cuddly (or Sir if you like), But I'll be happy to tell my name to a nice sweet REL sub Female ☺ I'm just a nice, caring DOM who will treat you great when your good, but have no problem putting you over my knee when your bad. (But sometimes being bad is fun ☺) If your for real, and you'd like to meet a nice Dom guy, don't be afraid to IM me ☺ don't worry I don't bite (well not too hard anyways ☺ "To err is human, to forgive requires a spanking first" "Did you bring the handcuffs and the chocolate syrup?"

The below chatroom chat discusses how dominant males can have several submissives living with him, while doing so in a civilized manner. I would like to point out, while all of this sounds wonderful and innocent, in viewing the web sites on the D/S lifestyle, never have I witnessed such brutality given to another human being. The photos show welts, scars, bleeding, unbelievable restraints, and tortures that were unimaginable until witnessed. Even the marriage ceremony between a dominant/submissive ends with the submissive being abused and tortured.

Investigate and think clearly before venturing into this lifestyle—for it may be impossible to escape.

A CHATROOM CHAT:

Lady Jet:	by then you hope to have your commune organized R/T?
R/T Trainer:	one slave is the domestic slave
R/T Trainer:	the other two, plus me, work
DomWhoCares:	That is a good plan
Dovie:	I need a domestic slave very badly
R/T Trainer:	the domestic slave gets free room and board and a stipend
Lady Jet:	R/T...and what qualifications does the domestic slave have to have?
Learn'n Sub:	R/T...do you seriously think that domestic stuff is NOT work???
DomWhoCares:	Stipend?

R/T Trainer:	Sub...keep reading
LadyKitten:	you do Dovie???? I guess Dovie isnt doing a good job hehehe >ducking<
R/T Trainer:	of course it's work
DomWhoCares:	The work should be her honor and pleasure
Learn'n Sub:	free room and board that she has to clean and cook
R/T Trainer:	I meant a commercial job for the other three of us
Learn'n Sub:	WhoCares...good luck!
Dovie:	No Kitten, she isn't.
Lady Jet:	R/T...what about retirement for the domestic slave...the worker bees will get Social Security
R/T Trainer:	you missed the stipend part, Lady Jet
DomWhoCares:	Yes I know
R/T Trainer:	stipend + medical + retirement
DomWhoCares:	Don't forget fringe benefits
SoftHandedDom:	401K for the domestic slave...or maybe stock options <g>
Lady Jet:	ah...ok...
DomWhoCares:	A day off
DomWhoCares:	vacations
DomWhoCares:	Paid of course
Lady Jet:	<—signing up for domestic slave job!!!...am tired of the workday world!
Lady Jet:	lol
R/T Trainer:	LOL
R/T Trainer:	good girl, Jet
R/T Trainer:	please wait until I move to Bloomington, IN
R/T Trainer:	and buy a large house
Old Dom:	Grins...Jet is about to retire, way to go Jet
R/T Trainer:	with a basement ;-)
DomWhoCares:	Actually I think it is doms who should be paid!
Lady Jet:	R/T...FORGET it now...i lived in IN for 2 years...i am not moving back to the
Lady Jet:	midwest
R/T Trainer:	yes, DomWhoCares, that arrangement is traditionally known as "pimping"
Old Dom:	Yeah, all those incipient Dom Elbows and such
DomWhoCares:	We have to constantly think of tasks for our slaves to perform
DomWhoCares:	And ways to have them please us!

R/T Trainer:	well Jet, I have never lived in the midwest, so of course I can idealize it <smile>
Lady Jet:	Hey...retirement looks good on a day when you are tired of work world!
DomWhoCares:	Well then wouldn't the paid slave then be considered "whoring"?
LadyKitten:	how come the slaves dont think of things to do on their own to pleasure you WhoCares :))
R/T Trainer:	and you didn't live in a D/s commune complete with dungeon either!
DomWhoCares:	That is a very good question Kitten

NAME: WillTrain, **CITY/STATE:** San Francisco, CA, **STATS:** Male, Single, **DESCRIPTION:** Sadist to the female who needs it. Transformer of the woman who wants it. Owner of the slave who craves it. A powerful woman breaks wearing her pride. She is transformed as her Master strips her of it one layer at a time. Finally, she wears her collar and leash with pride and serenity. If your needs, too are service, sexuality and slavery, inquire withing or contact my partner and beloved slave, XXXXXXX, to join us. Add to your lifestyle; Don't replace it.

NAME: LetMeTrainU, **CITY/STATE:** Chicago Metro Area, **STATS:** Male, **DESCRIPTION:** Will initiate you into the world of D/s and B/D. Subs and Slaves will be considered for R/T only. Taking the raw talent of a would be sub and molding her into exactly as I desire her to be. You may mail Me if you are not presently collared or training with a Dom now. I am a sadist...I have limited time so only serious will be considered. You may straddle my PC if you desire...but we'll usually tan with floggers...welt with switches, decorate with clamps...need more? I don't want to be your boyfriend or husband and I prefer yours stays out of my business...I am stern...controlling...many rules...but always with a touch of love and respect for who the individual sub is...and if your true sub/slave material...you'll thank me on a daily basis. Currently accepting applications for a bi sub sister to My Little Pet.

NAME: CallMeTrainer, **CITY/STATE:** Venice, FL, **STATS:** Male, **DESCRIPTION:** Training School Fully Equipt Dungeon-Restraint& Pleasure-Pony-K9 Real Time only- Special Attention to Mother Daughter -Sister teams. Fetish Equiptment-Adult Toys-Adult Films (www.XXXXX.com) Fetish Club(www.XXXXX.com) IF HE WONT I WILL!!

I have saved what I consider the best-of-the-best of the dominant male profiles for last. There was no name, city/state, or stats associated with this profile, just words written by a wannabe poet/writer, or someone who is extremely sadistic. May you enjoy these penned words for what they are worth.

PROFILE: She asked me "Master, can you take away my pain?" No, I said. It is not for me to take it away but for you to give it up. She asked me "Master, can you take away my weak thoughts?" No, I said. Your spirit is whole, your thoughts are only temporary. She asked me "Master, can you grant me patience?" No, I said. Patience is a by-product of tribulations, earned, not granted. She asked me "Master can you give me happiness?" No, I said. I give you the tools, happiness is up to you. She asked me "Master, will you make my spirit grow?" No, I said. Your spirit must grow on it's own, but I will prune you and help you grow. She asked me "Master, may I have things so that I may enjoy life?" No, I said. I will teach you to live life so that you may enjoy all things. She asked me "Master, will you teach me to love others as much as Master love me?" I said... Ahhhh...finally you have the idea. I will reward you for your effort to understand and grow. Two hours later, I released the ropes that bound her to my ceiling, her exhausted body red and tender lay motionless on the floor. She thought to herself..."I must be the luckiest woman alive."

This lifestyle is violent, even though dominators and submissives would greatly disagree with me, preferring to view it as a showing of the greatest love and trust one can have for another individual. The intimacy in these relationships seems to border on the sadistic side of life in order for one to achieve a close intimate relationship with another. In both the dominator/submissive lifestyles, no matter if the dominator is male or female, the submissive must always give up their power in order to prove their love and trust for the dominator.

This is a lifestyle many people practice on a daily basis, and one that actually seems to work for them. I am not referring to the "wannabes" who would like to try it on a boring Saturday night, but to the thousands of people who live this lifestyle everyday. These people seem to have found a lifestyle that compliments and coincides with their very daily existence, even though the "vanillas" of the world (which I admit I am one), cannot possibly relate.

UNUSUAL, RAUNCH, PIGS

In this chapter I will let you be your own personal judge on these profiles, and keep my innermost feelings to myself (this is a very hard task for me to do in this chapter).

Being once naive, I really had no idea such lifestyle preferences existed for men, and I am hoping now they are nothing more than mere fantasies in someone's mind that has been brought to life in a public chatroom. But, if these lifestyles are not a fantasy why would any man want to write about it and put a name to it for the entire world to see? Could these men just be looking for the shock value in the written word? I can only wonder, as here I have found words and requests that I never dreamed could exist from men that I could possibly know in the real world. Their perversions are such that they must use a color code system (see Hankies in appendix), which defines their sexual preferences to like-minded men since private Internet providers do not allow blatant perversion in their chatrooms. This color code system is also used extensively in the gay male world for hooking-up with others who have the same likes and dislikes in the bedroom, in the kitchen, or wherever.

Unfortunately, it is impossible for Internet providers to censor every profile everyday, and no Internet provider should be personally blamed for these profiles, but blame should be placed where it belongs, with the profile creator for creating such a profile that anyone, including children, can have access to. What you, as a responsible adult, can do when you run across a profile that is blatant and degrading is report it to your Internet provider, who will then take action against the creator of the profile.

I have chosen not to list any chatroom chats from these men, preferring to list only their profiles, instead. Here now are, as they call themselves, "The Pigs".

NAME: BabyBoy, **CITY/STATE:** USA/International Travel—Based in NY, **STATS:** 23 YR old, male, 5'9", 175lbs 44ch 16a ripped, **DESCRIPTION:** No one over 40…dont really need the biz that bad. 6%bf work out not a BB 8c pierced, tatts not the boy next door, i'm a smelly nasty rauchy sleezy sick motha f**kr I'm white street trash been hustlin since i been 14 luv the job u must be into: ruff sex, drinking, drugs, smokin cuz that's who i am who wants some pretty boy doin it to get thru collegs. i do it becuz i'm a pig an luv it. My scenes r dirty-nasty-sick&twisted I will do anything NO LIMITS (sever beating, SM, WS, FF, puke, skat, raw, knifeplay, gunplay, body modification & mutilation, burning, cutting, blood, cum, snot, spit, K9snuff. I aint here 4 idlechat/trade/cyber. Serious only into above. Have 2 pics serious only get

This next profile has given me pause for riding in the elevator at the office. As it happens I work with a lot of suits in a very professional setting and now this little profile has put a whole new prospective on my elevator riding. "Oh, please tell me it is very expensive cologne I am smelling!"

NAME: PiggyInPen, **CITY/STATE:** L.A., **STATS:** Male, short, dark, hairy, handsome, 40, **DESCRIPTION:** nasty under the surface i like corporate, nerd, suit n tie type of guys (or501s) who might wear days-old jockeys n dirty jocks under their suits or jeans. We see each other, elevator/ party/ street, look into each other's eyes, and you pick up that it might turn me on to be a pig I am totally turned on to rank armpits, dirty sox. dirty Jockeys / jocks ... being pissed ... a ripe hole rubbed ovr my face Hey man, whatcha got on under that suit or 501s.

NAME: Mr. Topper, **CITY/STATE:** NJ across from Nyc...easy access to city, **STATS:** 55YRS, youthful in shape guy, **DESCRIPTION:** controlling pigs into eating sh..., drinking piss... also like pigs that like man stink and like to lick and sniff it... using pigs for pleasure... you like it boy

NAME: Animal4U, **CITY/STATE:** San Francisco Bay Area, CA, **STATS:** 36, male, **DESCRIPTION:** extremely submissive bi zoo male. I can be extremely generou$ for a Master or Mistress who can use me with their K9. You don't have to participate if you don't want to. I'm experienced and very very serious. Open to other farm as well, if you like Age/race in unimportant, attitude is. I'd also be interested in making a gay farm movie, just use, you keep the money. "four legs good, two legs bad" Orwell

I was totally unaware there was such a thing as an "erotic" enema until I came upon this profile. Having had several surgeries in my lifetime, the thought to even associate an enema with pleasure, never entered my mind. I do believe this fellow could use a good nurse in his life!

NAME: FillMeUp, **CITY/STATE:** Southern California, **STATS:** mid 40, male, div, **DESCRIPTION:** I love enemas... All kinds. Bardex, colon tubes, bulbs... enema...enemas...I love em. Bardex, colon tubes, bulbs. Looking for nice white female to give me erotic enemas Will give same in return. Am very discreet...married OK Looking for real time only in S. Calif. And I really love prostate massages too...and love regular great sex too...A LOT.

While not finding the following profile to belong exactly in the "pig" category, I did find it particularly distinct from other profiles, and felt a need to include it. I believe once long ago I read a short story that portrayed such a man, and of course, once again I dismissed it as being absurd. Now I am eating my words, again! Ok, I accept this man's request, but I really want to know where one finds a woman with a baby who has time for this?

NAME: CowMan, **CITY/STATE:** SF, CA, **STATS:** 22, male, 6'1" brn hair, bl eyes, 190lbs, single/available, **DESCRIPTION:** Searching for a pregnant/ lactating woman.—I'm not looking for ONLY suckling, I want all the fun! ☺ Enjoying the pleasures of lactation—flirting, pleasuring, enjoying pregnant and lactating ladies, inducing if desired. Got Milk?…for me? <wink>

NAME: Lookin4it, **CITY/STATE:** Portland, OR, **STATS:** male 40, **DESCRIPTION:** pigbear looking for a slave brother. Looking for part time Master (s)—into a slave w/mind and experience- also MUST Be into Raunch!! i am an Ex-Top and Proud of it! scissors, rock and paper, employed www.XXXX.com Orange hanky kinda guy- into s/m, b/d, cigars, ws, va, ect... - "Learn the words the Internet won't let you use"—cigars, rubber, leather, uniforms, wrestling, training, confinement, electricty. Really getting into raunch-what else is there to say but i am a proud sla

NAME: MasterLeather, **CITY/STATE:** Syracuse NY, will travel N'east for REAL encounters, **STATS:** 6'2", 215#, 54, 'stache/ goatee, silver-brown long hair, size 13 feet, gay Married, **DESCRIPTION:** Leathermaster, Hot, Healthy, Leather, Brown Hankey LEFT plus: Black, Red, Yellow, Beige, Gray, Dk. Blue, Heavy in tense, real, & committed. 25 yrs. Of experience. OBEYS!...as does my branded, tattooed & collared 24/7slave. Artist/photographer…Dominant control & raunch training. Wilderness & edge trips. Honesty counts. Creative action. The only true perversion is never letting the beast within out to play.

NAME: Lovin It, **CITY/STATE:** Washington DC area, **STATS:** 41, male, single, d/d free, safe and sane, **DESCRIPTION:** Toilet, GS, BS, CBT, Toys, Golden showers, Kink, Submissive, Dominant, Switch, anal, oral FF, butt, plugs, feeder, piss drink Fun, wild, kinky SWM ISO kinky women. Into receiving golden. Love to drink, love to take big toys. Love to orally please. Into pure raunch and kink!

Remember, if you need help with any of the abbreviations contained in these profiles, there is an appendix in the back of the book for your use. In doing my research I would read a profile then navigate to a web site that could explain what I had just read. Hopefully, I have made this a much simpler process for you. Also, in the glossary you will find web links on the various lifestyles you may want to reference for yourselves.

NAME: Piggy Ride, **CITY/STATE:** Western NY, **STATS:** 35 years old, male, 6'2", 195 lb, full beard, brn/blue, **DESCRIPTION:** Into raunch, red, beige, tan, yellow, brown, gray, some black, tt, cbt, sounds, light pink dark pink, more just ask love cigar/pipemen and get into forced cigar/pipe smoke, can be mutual on occasion.

NAME: Gross Out, **CITY/STATE:** North Park, San diego, **STATS:** 41, male, available, **DESCRIPTION:** raunch, sniffing, licking, pitts, socks, jocks, cbt, man stuff...yes Sir!

NAME: RAUNCHY, **CITY/STATE** North Jersey, **STATS:** 35, GWM, 6-1, 185, 8.5 cut, horny, **DESCRIPTION:** Pit raunch, sniffing sweaty hairy pits on clean cut guys, poppers Love J?O, sucking hairy cock. Married/partnered buzz off. No pic, don't mail me. NOT into other raunch. If you smell and it gives you a boner, mail me NOW. If I'm on, I reak and I'm looking for other guys who get off on their own pits.

NAME: HearMeStraight, **CITY/STATE:** Midwest, **STATS:** 5'11", 180, 45c, 34w, 16 biceps, 40 yrs old, male, **DESCRIPTION:** Kink, leather, etc. Very chem friendly... (If your life needs 12 steps, keep the f**k away from me!) Keywords: leather, ws, ff, raunch, SM, raw, BB...PROUD BOTTOM, looking for Leather Tops to age 45. During the day, I have a high profile job...so no facepics here. Deal. No profile, NO chat... No Pic, NO play...NO beginners! POWER EXCHANGE IS THE ULTIMATE SEXUAL HIGH

NAME: Mr. Piggy2U, **CITY/STATE:** L.A., O.C., Long Beach, **STATS:** 34 yr old, male, single, 5' 11", 170lbs, buzzed, blue, tatts, goatee, clipped, 32 waist, nice body, 8"thick, **DESCRIPTION:** Leather, nasty, bareback, hardcore, cum, kink, party, wild, intense, uninhibited, boots, buttplay, slam, FF, tatts, piss, uniforms, pigs, raunch, groups, sweat, slings, role play, sleaze, buttplay, versital, rough, chem-friendly, playroom. If you're a twisted party pig and you wanna play nasty, mail me!

NAME: PrettyBoy, **CITY/STATE:** Arlington, VA (DC, NOVA), **STATS:** 53 GWM, single, 6'1, 180#, blue eyed, **DESCRIPTION:** ISO nasty pigmen; raunch, hankies: yellow, beige, red, blues, white, brown, grey, and now pink! Sleazy Leather pig! Kink; chem friendly; have bushy white-moustache/beard. Hey, i want in there too, OINK!!

NAME: TexasFun, **CITY/STATE:** Dallas, TX, **STATS:** 40 yr old, male, single, 6'1, professional, **DESCRIPTION:** Golden Shower fun, females only for wet fun and kinky activities. "Rather be pissed on than pissed off."

NAME: BigB, **CITY/STATE:** Long Beach, CA, **STATS:** Male, single, Gay, 6' 2", 220 HIV neg, **DESCRIPTION:** toilet training by agressive top or dirty mutual times with raunch buddies. W/s, scat, humiliation, discipline, kink. Online & ready, seek only guys really into it. looks unimportant attitude is.

NAME: 2forSmoke'n, **CITY/STATE:** South Florida, **STATS:** Dad is 5'9 185, buzz salt & pepper, stach, blue eyes. Boy 6' 200, buzz drk/drk stach, **DESCRIPTION:** DadS and I am His boytoy...if i am on WE are looking to play! My cigar leather Dad and I are looking for a 3rd, 4th, etc for Hot Cigar Leather Play. Hot for LEATHER, CIGARS, Tattoos, Cops, Bikers, Daddy's and their boys, w/s, raunch, role play, Kidnap, Fantasy. Take this boy, and let Dad sit back, with His Cigar and watch You smoke Your Cigar and use His boy... Likes to party. Bring Your boy and let Dad play with him. BD/Hoods/gags/restraint, LEATHER and CIGARS a MUST. Forced Smoke, tight gloves for the boy...looking for HOT wild kinky, dirty fun...Want to know more, just ask...

NAME: Good Lick, **CITY/STATE:** San Francisco, CA, **STATS:** 36, male, single, 5'6, 150, br/br, shaved head, stache, **DESCRIPTION:** cigars, boots, gloves, leather, hoods, gags, rubber, raunch, sleaze, spit, pigs, getting rimmed. Just ask prefer to be on top or mutual. Pierced, tattooed. No drugs. Alcohol/poppers OK

There seems to be a lot of showers being taken by this group of men, but I don't think it's the type of shower most of us are taking in the morning. One could venture a quick guess from these profiles that these men are into the raunchier side of life, and really get their kicks from urine and feces. Why? I have absolutely no idea, as I cannot even call it any type of sexual preference, since it is not what I could call a sexual act. It is outright filth! If I'm being too judgmental, I do not apologize.

NAME: LuvsGolden, **CITY/STATE:** milwaukee/wisconsin, **STATS:** 36 yrs old, male, single, **DESCRIPTION:** golden showers maybe brown Looking for beautiful women to give me a golden brown shower. Willing to pay top price. Onlt the very attractive need apply.

The next three profiles are our baby men. Can you all picture these big, hairy guys crawling around on the floor, talking baby talk, while enjoying the comfort of diapers? And, just who the heck is changing these diapers? I cannot

imagine any normal (not quite sure what normal means anymore) woman actually enjoying being in this lifestyle. It was bad enough with a child, let alone a grown man!

NAME: Babycakes, **CITY/STATE:** Mid-Atlantic/East coast/ southeast, **STATS:** 30 yrs old, GWM, single, diapered, **DESCRIPTION:** men, diapers, men in diapers. I like to wet and mess my pants and diapers...would like to meet others who like the same; interested in friendship and LTR in diapers Q: "Do you wear boxers or briefs?" A: "Depends."

NAME: Baby2Diaper, **CITY/STATE:** Southern, CA, **STATS:** 38, male, **DESCRIPTION:** boyish men in diapers (prefer 28-40), interesting conversation, darts, camping, movies, bike riding Please have a PROFILE to chat. NO cyber or phone.... PLEASE don't ask!

NAME: Lit'l DiDe, **CITY/STATE:** Babyland?, **STATS:** Male, Adult Baby, **DESCRIPTION:** Being treated like a baby by females, Wearing diapers, Roleplaying, Sucking my thumb, Pacifiers, Just being a baby. Mommy are you in here? Adults only Please.

NAME: Mr. Shine, **CITY/STATE:** San Francisco Bay Area and Marin County, **STATS:** 6' 220 blonde, blue, big bushy beard, **DESCRIPTION:** CHROME COCKRING luvr, vanilla sucks, I got BOOTS, WS too TITPIG still a PIG, betcha UR2, gruntin', sniffin', snortin', stickin' HARD! In HOTPIG holes. POWER SLAM RAM version-→ 7.5 long X 7 middle X 8 at base, UC raunch pig fillin' potholes with my fat UC BEERCAN. I have big bushy beard, tatts, pierced nips with 2 guage circular barbells, raunchy JOCKS & UNDERWEARS get me goin' too

NAME: KissesButt, **CITY/STATE:** Midwest, USA, **STATS:** 42, male, 5'10" 197, **DESCRIPTION:** MEN, beards, manfur, boots, cigars. I dig giving TOTAL service. Like lotsa WS and VA, raunch, humiliation, domination, control, toilet training, force feedings, scat, spit, snot, whatever. Love Masculine MEN who KNOW how to treat a pig. Twisted, groveling pig for a MAN. (healthy, masculine, professional by day) Take me LOW, where I deserve to GO, whether I like it, OR NOT! "CUFF me, ROUGH me, then STUFF me!"

NAME: Mr. Swine, **CITY/STATE:** Tampa Bay Area, Florida, **STATS:** 40, male, partnered open relationship, **DESCRIPTION:** Hankies (all on right) Black, Grey, Yellow, Pale Yellow, Teal, Red, Blue, Lt. Pink, Dk. Pink, and many others. Total pig bottom into anything that involves kink and raunch, the nastier the better. Member Watersports Enthusiast of Tampa Bay. Other interests include saline infusions, sounds, catheters, vac pumping, leather, jockstraps, restraints, hoods, masks, boots, dirty white socks, rimming, 3+ ways.

NAME: TLKDIRTY2ME, **CITY/STATE:** upstate, NY, **STATS:** 31, male, 190, 6', **DESCRIPTION:** NEED DIRTY TALK NOW... WANNA MEET (I MEAN DO) YOU WHEN I'M IN YOUR TOWN. Into DIRTY talk with like minded individuals. NOTHING is too sick and twisted for this pig. Into EXTREME chat about raunch, LOVE BAREBACK/INFECTION talk, group events, forced events, totally anonymous sex, one night stands, bathhouses, gloryholes, bookstore. HAIRY HUNG, DEEP OPEN MIND, More interested in talking with VERSTILE GUYS PLEASE NO*** REPEAT*** NO TOTAL BOTTOMS! I like 50/50 scenes, I get messy and used and you get my filth and funk in return. Telling me what gets you HARD will get you further than "whats up" or "nice profile"

NAME: ValleyPigBoy, **CITY/STATE:** Los Angeles Valley, CA USA la sfv CA LA, **STATS:** single, GAY, brn brn 5'10 185#, **DESCRIPTION:** VA CBT WS raunch mud sweat men versitile swallow use sneakers boots socks jocks uniforms degrade humiliate role play rim *sscrack beg boy training biker leather service lick sticky woof dog badboy force flogging measured pain wrestle for top overpower sticky hard drive that produces lots of swimmers made to travel deep, changes floppy to hard drive in a few (key) strokes Office Mgr. By day, PIG remaining 16 hours. If it smells sniff it again feet pit stink ripe crust crud slime OINK. What was your name again??? slurp gulp cum bark. Make it happen. Doesn't matter who's on top as long as we all get it in the end <grin>

You have just ventured through what I call the worst of Internet chatroom profiles. Maybe I am being too hard and too judgmental on this group of men, but I doubt it. I cannot imagine any of these men being someone involved in my daily routine, but they very well may be. And, in your daily lives as well. This is just a sampling of this preferred lifestyle from the Internet chatrooms, for there are thousands more just like them out there.

Who are these men? Where did they come from? And, the one question that I would like answered, "How did they get to this place in their lives?"

THE WOMEN

BI, GAY, LOOK'N

Bi and gay women differ in a chatroom from the men in this same category in that they most always insist you have a profile before they will speak with you. They are also more outspoken, including downright insulting, to others who have wandered into their realm without a profile. Being verbal, as if on the defensive, seemed quite common in these Internet chatrooms. I'm not sure if I equate it to a chip on their shoulder or the fact that these women are not too accepting of other people who differ from their lifestyle. It is almost as if they take the defensive with a newcomer instead of offering friendship and getting to know the newcomer first. And, they are not particularly kind to voyeurs. Yes, it is definitely a verbal cultural for the gay woman, particularly the butch female.

This is where I ran into the most vocal of the gay women, and I would equate her tolerance of other people to that of a bull that has staked his claim to the herd, and no one had better enter his pasture. In chatrooms I found most butch women to be rude, aggressive, and downright abusive to anyone they thought did not go along with their train of thought or had entered their chatroom conversations without being invited.

While a gay male will be extremely blunt and straightforward when it comes to sex, the gay woman doesn't seem to respond quite so openly in a public chatroom. She seems to prefer the comfort of a private chatroom for her sexual escapades. I must note here though, that I did receive a few invitations to private parties in individual's homes, which also included invitations to other members of the chatroom. Some invites were for a weekend stay of partying and fun, while others were for a single night. I never accepted any invitations, so I can go no further with what transpires at a gay female weekend party, but I would surmise the computer is probably not involved.

Several of the bi/gay chatrooms for women that I visited were void of conversation, except for the new chatroom arrivals entering and giving a general "hello" to anyone who would answer. This tends to make me believe that most of the chatroom occupants, as the chatrooms were to full capacity, had hooked-up with a partner and were buried in private conversations with each other. I did stumble into a bi chatroom that had quite explicit sex taking place with everyone and anyone in the chatroom, and I have included that chatroom chat in this chapter.

The chatrooms I did find that had chats taking place were often very enlightening, at least to myself, on the life of a bi or gay woman. One thing I need to point out is the fact that during my research most of the women's profiles were very short and to the point. That leaves me with the thought that maybe it's not really the women who do all the talking after all, but the men. The men's profiles could ramble on for pages, but the women had very little to say, as you will note, but what they did say was straightforward and to the point. Hmmmmm, could it be that men have given women a bum rap all these years in order to pass the buck?

Here now are some of those chats and profiles from the Bi/Gay Chatrooms of women.

A CHATROOM CHAT:

CutieBabe:	hmmm... i'm in pa Couple
Couple4Bi:	hi
Debbie249:	28/bi/f/pic s2r
Couple4Bi:	what part of pa?
Call4Chat:	33/f in Chicago burbs, anyone local?
CutieBabe:	south east... or north of philly easier to say
Talker98:	right here, Call4
Couple4Bi:	kool
SoBelle:	i am in sw burbs
Couple4Bi:	u into cpls?
NeedLots:	Hi all
Call4Chat:	hiya Talker and SoBelle... western burbs here
LoverofBi:	hello GA woman here for someone
Call4Chat:	city or burbs Talker?
NeedLots:	24bif looking for same in cal
CutieBabe:	to be honest... never been with a couple 'yet' *wg*
SoBelle:	i read your profile Call4, i might be to kiddie for u
LoverofBi:	anyone here also from GA, u wouldn't be disappointed
Call4Chat:	ooh, how old SoBelle?
WantsU:	hello all
SoBelle:	21
CutieBabe:	hiya WantsU
OhOhOh:	o°°·•¤couple no one is intersted¤•·°°o
Couple4Bi:	well we r haveing a party this weekend there will be 2 bifems there we want more
Couple4Bi:	dont give up hun

WantsU:	18/very very horny f with sexy self pics im me to trade/ roleplay anything goes
OhOhOh:	o°°·•¤have D bring some of the girls she works
OhOhOh:	with¤•·°°o
OhOhOh:	o°°·•¤:o)¤•·°°o
Couple4Bi:	Cutie u waana CUM to a party
CutieBabe:	what part of jersey are ya in
Couple4Bi:	yes we can go there thats right i 4got
Goodie:	nyone in Mass.?
OhOhOh:	o°°·•¤Ocean County¤•·°°o
Couple4Bi:	45 min from u
BUTCHIE:	hard butch here
CutieBabe:	not too far
Couple4Bi:	think about it
OhOhOh:	o°°·•¤what about getting girls to cum
OhOhOh:	here???¤•·°°o
Couple4Bi:	its going to be fun
BARB4U:	<~~~~ 32/BIF/ Virginia
WantsU:	anyone want to help me out and talk to me
OhOhOh:	o°°·•¤oh yesss lot of fun?¤•·°°o
WantsU:	very very lonely and very very horny f with sexy self pics
BiGirl:	26BiF in Boston looking for either phone / real meeting?
WantsU:	if so mail me
CutieBabe:	awww well where is here OhOh?
PlsCall:	we are looking for openminded females, please see profile
OhOhOh:	o°°·•¤Ocean County¤•·°°o
OhOhOh:	o°°·•¤I am the one having the Party on Sat.¤•·°°o
CutieBabe:	oh duh! lol
Couple4Bi:	Cutie, Oh is my wifes girl friend
OhOhOh:	o°°·•¤yes we are¤•·°°o
PlsCall:	we are looking for openminded females, please see profile
CutieBabe:	I'd love to say yes but I have a play party on saturday night
CutieBabe:	:(

The following three chats are very short and very straight to the point in what the females are looking for. Profiles containing "NO MEN" are listed in many a gay female profile, and in the chatrooms I noted that men would often enter a chatroom and begin heckling the females. I never did see a female enter a gay male chatroom and begin heckling the men, though. Interesting that straight women seem to tolerate a gay male, but some straight males can't seem to tolerate gay females. Could this have anything to do with their own masculinity and not being able to conquer every female that crosses their path?

NAME: Cum2Me, **CITY/STATE:** east coast, **STATS:** female, married, **DESCRIPTION:** other females mmmmmmm, female chaser, cmere it will feel good!

NAME: LuvBoobs, **CITY/STATE:** here/there, **STATS:** female, single, **DESCRIPTION:** No SINGLE MEN!!!!!!!!!! I like large Breasts. I'm 34b. Send your pic I will return...Would love to experience an ebony treasure.

NAME: FamilyFun, **CITY/STATE:** GA, **STATS:** married, **DESCRIPTION:** Hobbies—a few that include family, curious? All we want are peace and quiet, so give us a piece and we'll be quiet!!

Females in a relationship with another female, who are looking for a third or even another couple comprised of women, are extremely common in the gay female profile. Maybe they equate it to a "slumber party" back in their childhood.

NAME: 2FOR1, **CITY/STATE:** Philly, PA, **STATS:** female, 46, partnered with bif, divorced, **DESCRIPTION:** sex, sex, and more sex, family fun, both in medical—NO MEN NEEDED. Send pic for mine- will not send first—if at first u don't succeed-try again-I'll cum

NAME: Dom4Fems, **CITY/STATE:** Southern Cali, **STATS:** female, **DESCRIPTION:** Remember my name, you'll be screaming it later. No I don't have a pic, so stop asking me. Leather, whips, chains, roleplay, jello, leather, whipped cream, walking around with my riding crop being a Domme Fem Fatal...forcing slaves to do my bidding. Two lovely playthings dressed in latex...mmmm I may be funny but I can beat the S**t outta you...Can I pet your kitty...If you quit moving the leather straps won't bite into your skin...Not looking 4 a male slave, so males don't IC me

NAME: Luv'er of women, **CITY/STATE:** here, **STATS:** female, **DESCRIPTION:** girls, girls, girls, I also love to be treated nice! Ladies come and get me hehe, No GUYS PLEASE!. That means you people who has the funny looking thing hanging in between your leggs, hehe. Loving and pleasing girls, women, ladies, females, and members of the my sex, got that guys, it means no males at all!!!. When I am good, I am bad, when I am bad, I am even better, but when I am wet!!!I am the best!!! Lol, IF YOU DON'T HAVE PROFILE PLEASE DON'T WASTE MY TIME AND YOURS, I WILL JUST IGNORE YOU!!!, ALSO LADIES, I LOVE YOU!

NAME: BeBe, **CITY/STATE:** Dallas area—Irving, **STATS:** female, married (I Love my Baby), Beautiful Ebony female with brown sugar skin, **DESCRIPTION:** We are looking for a bi-female in the dallas area to have som fun with. We play together. No sex with him, he just wants to touch, lick, kiss and play. We are a very professional, discrete, drug & disease free couple and expect the same. We are "virgins" at this so please be gentle. If you would like to hook up with us for some off line fun send us an IC…"NO MEN" he is STR8.

Here we have a very, very short profile, but I had to include it just to prove diapers are not for men only.

NAME: BabyGirl, **CITY/STATE:** PA, **STATS:** female, single, **DESCRIPTION:** diapers, babyplay and girls. Want to play baby with me!

NAME: StudentGirl, **CITY/STATE:** SMU TX, **STATS:** student, 21 yr old, female, **DESCRIPTION:** 3somes or more maybe, taking pics of young girls, I love to get high and kinky

NAME: CallMeButch, **CITY/STATE:** why? Coming to visit? **STATS:** female, single, **DESCRIPTION:** femmes, softball, rugby, traveling, camping, Gormet cooking and all those home shows on TLC! "Hard enough to be a man but without the weenus problem"

Below is a chatroom chat, which contains a female who is a butch, and one male heckler who the chatroom occupants are avoiding. Also, it seems by the end of this conversation our butch female may have scored. Just like a night out in the local town tavern—all the girls look better at closing time.

A CHATROOM CHAT:

BadButch:	I've fucked women size 2 to size 24
MaleMan:	NEED A LIL MALE LOVIN
Kind&Ready:	looking around
BikiniGal:	ok badbutch, i got your point
Quietstrrm:	bad…next time i'm out, meet me for a drink if you can…since we live so close
Kind&Ready:	did the threesome Domme leave?

BadButch:	any size can be sexy as hell
MaleMan:	<<<— MALE HERE
Baby Doll:	not if you were made of gold with a diamond in the tip of it
Quietstrrm:	but i'm usually at Woody's or Uncles
BadButch:	I snap lil freaky guys like you...easy as a chicken bone
BikiniGal:	i guess i should have never brought up the weight of those two butches
Jenny4Ever:	Come to the confused couch Willing u can sit on my lap
IN'NBIF:	walks over to Bikini...
MaleMan:	WHO WANNA SIT ON MY LAP
IN'NBIF:	want a drink girl
BikiniGal:	WHOS GONNA SIT ON MY FACE?
WILLING2:	lol Bikini
IN'NBIF:	i will grins
BadButch:	never been to Uncles
IN'NBIF:	hmm Bikini rotations...
Jenny4Ever:	holding Willing on my lap...Willing meet Kind
WILLING2:	hi Kind
MaleMan:	ALL YALL CONFUSED
Jenny4Ever:	and Bikini
MaleMan:	GET A DICK
Kind&Ready:	hello Willing...smiles
WILLING2:	hey again Bikini
BadButch:	I have been to the Bike Stop and that horrbile Sisters place
Quietstrrm:	i usually hang out with the guys, instead of the "I'm pretending i'm not such a hard up
IN'NBIF:	oh go away thin small man
Quietstrrm:	lesbian bunch" lol
Cindy Lee:	oh god lmao not to respond to MaleMan
Jenny4Ever:	and of course the lovely Ms BadButch
BadButch:	lol
Quietstrrm:	Male
Kind&Ready:	smiles
BadButch:	lovely??
WILLING2:	ms badbutch (kissing her hand))))0
Quietstrrm:	darling...but i have one
OhSoPretty:	giggles at Erica
MaleMan:	ERICA

Quietstrrm:	of the latex variety
IN'NBIF:	so Bad butch...i dont believe we have met...extends my hand to you
Jenny4Ever:	ok, the badbutch Ms badbutch
Kind&Ready:	giggles
IN'NBIF:	im in'nbif
BadButch:	<actin all femme in my combat boots>
MaleMan:	YOU HAVE WHAT
Quietstrrm:	and baby, because its a woman fucking me...it feels so much better...lol
IN'NBIF:	are you a truly a bad butch? Smiles

As in the male gay world, the gay female world also contains the dominant/submissive relationship between two women. This seems to be a very common lifestyle, and one I am beginning to believe exists in all types of lifestyles, i.e. gay males, straight males, gay females, straight females, etc. Below are several of the gay female dominant/submissive profiles that are in search of partners for this venture into the pain side of life.

NAME: FDom4Female, **CITY/STATE:** Colorado, **STATS:** Female, Lesbian Domme, **DESCRIPTION:** Looking 4 sub f. Any information that you require I will speak of personally. If you are looking 4 BDSM, leather, zoophilia, blood, pain, urine, or feces look somewhere else. I'm very much into details and description. If you don't think you can take the necessary time, don't bother. If you are interested, mail me with what you look like, what you are wearing, and if you are owned.

NAME: Susan the Sub, **CITY/STATE:** New York, **STATS:** Female, Single & staying that way, mother to a 16 yr old daughter, **DESCRIPTION:** submissive to FEMALES ONLY. D/s with lots of bondage. Dressing sexy, Love lingire, nylons & heels, etc...

NAME: LetsPlay, **CITY/STATE:** East Coast of the US, **STATS:** Single LESBIAN, **DESCRIPTION:** Kinky sex, bondage, rollplaying forced scenes, degridation, S&M, D/S, pain (but no blood or cutting or burning) Ohh you wanted my vanilla side. I'm into REAL women (Key word is real) If you're a fake please just leave me alone. I work to pay for my Sex habbits (Sex toys aren't cheap you know!!!) Tie me up (or down) and force it hard...OK OK

NAME: ForU, **CITY/STATE:** FL, **STATS:** female, single, **DESCRIPTION:** Older women, K9s,,,,Fam Fun,,,, A lot of sex Send pic I promise I will send back (Please no men) Ladies Only

Donna Tracy

NAME: Lovin it2, **CITY/STATE:** Cincy, OH, **STATS:** single, female, student, **DESCRIPTION:** girls, girls, girls,…women in their 40s… round hole, round peg… HEY MEN GUESS WHAT I AM GAY… STOP GO AWAY I ENJOY WET STUFF, NOT HARD STUFF, soft kittens love to be tigers...MY HEART CAN BE WON WITH NAUGHTY PHOTOS...if your sending personals send 2 so that i know you are for real...death to all men on the net who pretend to women...

NAME: I'm A Sweetone, **CITY/STATE:** Tampa, FL, **STATS:** female, **DESCRIPTION:** Getting you to climax, bringing u towards my ecstasy, taking you so far into pleasure that your body will be aching for more, I will lead the way if your willing to follow… Full Time LoVeR GiRl (i'm a lesbian so men stay away!) I can take you there bring you high enough to see what heaven has to offer, and bring you down to earth and endulge ourselves in sinful pleasures and go for another round of sexing you down… (can you fell the tension?)

And below we have a chat where one girl insists on a goodnight kiss from her partner(?), and then proceeds to hit on someone else as soon as her partner leaves the chatroom. Kind of like the real thing in a boy/girl relationship that so many of us have experienced. I guess some things in life just aren't so different between the sexes after all.

A CHATROOM CHAT:

Sharon29:	yeah L—I been on that couch already...
LoisXXX:	come sit with me
Lets Talk:	I prefer the sofa
LoisXXX:	i won't hit on u i swear =)
Sharon29:	::::::blushing:::::::::
Sharon29:	been on the sofa too...
Lets Talk:	oh my
Chatty4Bi:	hi from 19 f
LoisXXX:	haha
LoisXXX:	poor lil innocent LetsTalk
LoisXXX:	forced to watch all the debauchary
Sharon29:	awww... L—I gotta go... like right now about
LoisXXX:	give me a good night kiss?
Sharon29:	have to get up early...
Lets Talk:	t6hats ok when it gets to much I just close my eyes
Sharon29:	goin over to L

Sharon29:	holdin your head—pressin my lips to your mouth...
LoisXXX:	;;holding her hand pulling her towards me gently::
Pic Collector:	hi i'm amanda 21 fem bi with nude pic mail me a pic to get it
LoisXXX:	::;closign my eyes and kissing her soft::
Sharon29:	mmm lingering...tasting...
LoisXXX:	slicking my tounge ring at her mouth
LoisXXX:	flicking*
Sharon29:	takin turns—with each others tongue
Sharon29:	mmmm love ur ring
LoisXXX:	mmm wish u could stay with me tonite love
Sharon29:	breaking away...smiling
LoisXXX:	but i will see u again =)
Sharon29:	time for me to get goin...
LoisXXX:	goodnight sweety
Sharon29:	yup... :-)
Sharon29:	see ya later Lets Talk...
Lets Talk:	bye
LoisXXX:	mmm slowed considerably
Bottle & Pen:	hello room 25/f with self pics
Lets Talk:	guess i will pour my self another drink
LoisXXX:	Letstalk...join me?
Lets Talk:	sure walking over to the couch
Lets Talk:	sitting down
LoisXXX:	::;making room for Lets::
LoisXXX:	::;raising glass::
LoisXXX:	cheers dear
Lets Talk:	cheers

NAME: Reptile, **CITY/STATE:** Texas, **STATS:** female, Divorced, **DESCRIPTION:** love to work out my long-warm-lizard tongue ;)~~~~~ u want something to feel good! Making You Beg For More, And Cum Hard... No Males Please, I have had enough of that...Females are my pleasure ☺~~~~

NAME: Cpl4U, **CITY/STATE:** Rossmoor, CA, **STATS:** 38DD, bi, sksbi fem, boyfriend, **DESCRIPTION:** I lovebig boobs. I love to play with girls, MEN DON'T MAIL ME I WON'T ANSWER, WON'T ONE WOMAN CUM WITH ME PLEZ, BEING ORAL, sucking licking, sucking licking Cple looking 4 bi fem to join us in pleasure, fun, & Extasy. PLAYING Looking 4 bi fem 25 to

45, to join us for immence pleasure. We're STDFREE. Llifes great, I WANNA SUCK ON IT!!!! PLEASE.

I had assumed, and once again I was proven wrong by assuming, that the below type of fetish belonged only to men. At any rate, I now believe it's possible for a woman to get her kicks from old, dirty sneakers too. But for the life of me I have absolutely no idea why. Is it possible that this could change the dress code of the frequent flyer?

NAME: High flyer, **CITY/STATE:** My Web Site XXXXXXX, **STATS:** female, single and bisexual, flight attendant, **DESCRIPTION:** I am so turned on by other girls in sneakers... I love to make girls with awesome sneakers my slaves... Family fun...and making OTHER girls do my K9... I am 100% dominate also... I also love to collect women's sneakers. Girls are sexiest in sneakers and socks only... While they submit to me 100%.... I NEVER SEND FIRST SO DON'T ASK.

NAME: Pretty1, **CITY/STATE:** Corn Country, **STATS:** Female, student, 34c-25-34, **DESCRIPTION:** want to explore with Ms. Right! Watching all the pretty girls at the local mall, deciding which one i will go after today. There is something about the touch and a kiss of another woman, that a man cannot duplicate no matter how hard he tries!!!

NAME: Serves2Please, **CITY/STATE:** Kneeling at the feet of my Mistress, **STATS:** Female, uncollared, **DESCRIPTION:** clips, whips, and my Mistress's leash...mmm...feeling my collar... pleasing my Mistress. A skirt and 5" heels and whatever Mistress alows. White slave serving my black Mistress, Kneeling before you, submitting myself to your desires and wearing your collar and leash.

NAME: Woman Lover, **CITY/STATE:** Amarillo, TX, **STATS:** female, married but i play, **DESCRIPTION:** NO MEN... i love woman, their touch, their smell, and the way they look, and i love to treat them like their the queen of the my castle. Being a bi-female... love to lick it ...let me lick it dry baby...let me in your heart and i will be forever there and you will never regret your did

This next profile I love! She is everything you would ever want in a woman, and all for the price of $200/hr. Heck, I think I could be "almost" anything you want me to be for $200 an hour, too! I'm also beginning to think I'm in the wrong business. I used to make $200 a week, before taxes, and thought that I was rolling in the dough! Oh, to be young again, and know what I know now— and, I hope you don't believe one word of that line ☺

NAME: BiLover, **CITY/STATE:** Richmond, VA., **STATS:** female, single, truly bi, 40DDD, **DESCRIPTION:** BIG breast and oral fun, 9 ways between my BIG boobs, I'm tall, slender, brown eyes, brunette/auburn, hot & do-able… All dates at my home $200 per hour…these 6" platform stillettos rule!!

Have I got a guy for the below bi female! She says she is lactating, and I just happen to know this guy from the previous chapter, remember him, who is desperate for a lactating female. Now, if I can just get one to call the other—who knows what will happen! I could finally be a matchmaker and go into the dating business!

NAME: JoinTheNavy, **CITY/STATE:** California, Europe, Canada, **STATS:** 36ee-24-36, female, Naval Officer, **DESCRIPTION:** Im A Bif (Dont Ask Dont Tell) Domme, I Sell Video's In Vhs And Pal Formatts, I also Have Exlusive Dvd Made. I Also Frequent Glory Holes. This One I Have 16000+ Avi's, Mpg's And Other Wonderful Pixs. I Am Lactating. I Am A Glory Hole Freak, I Also Love K9-Equines. Have 3.0 Send To Receive (s2r) Topless Only, Please No Men. I Have Suckle Time Availible Mail Me Send As A Attachment.

The below chat was taken from a bi chatroom and I have changed the names to reflect only male or female genre in order to simplify who is with who, or who is doing who, as the case may be. One part of this tits and ass scenario that brought me to complete laughter is the conversation that transpires in the midst of these flailing body parts between two old friends, and one mother having to give up the computer so her daughter could play a computer game. There has got to be a connection between sex taking place on the family computer and kids playing a computer game, but for the life of me I am still trying to figure out what it is.

A CHATROOM CHAT:

BEVERLY:	your tounge is so wanderfull
Jackie:	ohhhhhhhhhhhhhhhhhhhh yeahhhhhhhhhhhh
Sonia:	driving my fingers deep into Beverly's pussy
Steven:	slowly strokin my cock as i enter the room
David:	slowly driving the shaft deep
Brian:	mmm suck it baby while you finger my balls and ass
Jackie:	yeah deeper
Sonia:	sucking and nibbling on Bev's clit
David:	slow long stroke

71

Debbie:	OH YA AND I LICK YOUR BALLS TOO BABY
Gina:	::looking over at sonia and bev::
Jackie:	moving with your baody
BEVERLY:	rocking my hips to grind into your face and fingers
Sonia:	teasing her clit
Steven:	lookin arund hte room to see if any men need some servicing
BEVERLY:	looking up at gina
David:	shoving it in til your pussy swallows it all
Sonia:	withdrawing my fingers from your pussy
Jackie:	yesssssssss
Gina:	:::smiling:::
Sonia:	wb, Gina
Jackie:	thats what i wanted
Gina:	thanks Sonia
Sonia:	my tongue darting into your pussy
David:	my hips slowly rocking
Jackie:	Gina have fun
Sonia:	tasting your sweet juices
Jackie:	mmmm rocking with you David
Gina:	My pussY?
BEVERLY:	moaaning loudly
Cindy:	((((((((((((LULU))))))))))))
Sonia:	hands on your ass
LuLu:	omggggg, CINDYYYYYYYYYYYYYYYYYYYYYYYYY
LuLu:	where you been girl?
David:	fucking Jackie nice and slow
David:	savoring her tight pussy
Cindy:	yesssss lol great and you?*******
LuLu:	greatttttttttttttttt
Jackie:	mmmmmm yeaaaa
LuLu:	so good to see you
Sonia:	wet fingers probing your ass
Cindy:	same here...
Jackie:	aking mt pussy tighten up aroound your hadr cock
Cindy:	been wayyyy to long
LuLu:	Iknow, I know

72

Sonia:	my tongue plunging deep into Bev's pussy
David:	feeling it grip me
BEVERLY:	ohhhhhhhhhhhh soniaaaaaaaaa
David:	god your good Jackie
Gina:	::watching and playing with myself::
Sonia:	finger probing her ass
Cindy:	Hows pete ever see him??
Jackie:	mmm yea thats it
LuLu:	nah, I dont look, lol
Brian:	mmmm wanting to fuck Debbie in her tight pussy since she's gotten me so hard
Cindy:	hehehe
LuLu:	he's been off my b/l for ages, lol
Sonia:	sliding my finger in gently
Jackie:	mmmm ty your pretty good to david
Cindy:	ohhhhh mine to lol
Gina:	sucking on my tits
David:	rocking harder
Jackie:	i'm not far from showing you
Steven:	sittin back and slowly strokin as i watch the room
Sonia:	fucking Bev with my tongue and finger
LuLu:	well, my daughter wants on to play some games
LuLu:	was great seeing you
LuLu:	bye for now
Brian:	lay bakc baby so i can lick your wet pussy so i can slide my had cock inside
Cindy:	okkkkk same here hugggggggggggs
David:	pumping your pussy
BEVERLY:	pumping my hips deeper into your face
David:	filling it with my cock
Brian:	sliding my hard cock inside Debbie and pumping her pussy
Jackie:	yessssss yes
Steven:	watchin Brian pump his cock
Sonia:	looking u into Bev's eyes... cum for me, baby
BEVERLY:	my breathing gets deeper
Jackie:	mmmmmm yesssss
David:	my balls banging against your ass
Sonia:	holding you tight against my face
Jackie:	ohhhhhh ohhhhhhhhh

Brian:	mmm wish i had someone pumping me from behind while i pump her tight cunt
Jackie:	yeahhhhhhhhhhhhhhhh
Steven:	lickin my lips as i watch BRIAN
Debbie:	OH YES BRIAN HARDER BABY
Sonia:	fucking you hard with my tongue and finger
David:	harder and deeper
Gina:	::getting wetter::
David:	picking up speed
Jackie:	cummmmmmmmmmmmmmmmmm
BEVERLY:	mmmmmmmmmmmmmmmm yessssssssssssssssssss
Brian:	slapping my balls against your cunt and ass
Steven:	i get up and walk over to BRIAN
Greg:	::smiles:: hi Cindy
David:	feeling your pussy milk me
Brian:	fucking your hard
Jackie:	don't stop now go
Sonia:	feeling your pussy clench my tongue
David:	harder
Steven:	i kneel behind BRIAN
Jackie:	cummmmmmmmmmmmmmmmmmmmmmm
BEVERLY:	going to cummmmmmmmmmmmmmmmmmmm
Debbie:	hMM YES I FEEL IT BRIAN DON'T STOPPPPPP
Steven:	smack his ass
Cindy:	hello Greg
David:	pulling your hips onto my shaft
Gina:	ohhh
Sonia:	yeah, baby... let go
Steven:	and start to kiss it
Jackie:	mmmmm yes harder want you cum in me too
BEVERLY:	ohhhhhhhh yesssssssssssssssssssssssssssssssss
Sonia:	sucking your clit hard, Bev... biting it
Gina:	pussy throbbing
David:	my cock covered with your juices
Sonia:	feeling you cum wash over my face
BEVERLY:	looking at Gina panting harder
David:	pulling out...turning Jackie over
Jackie:	mmmm you feel ss ggod David
David:	on her tummy

Brian:	pumping deb harder and harder and reaching down to slide a finger into her tight ass
Jackie:	mmmm yea
Sonia:	beckoning to Gina
David:	moving in behind her
Jackie:	now what David
David:	grabbing her hips
Debbie:	OH YES, FINGER FUCK MY ASS BRIAN, I LOVE IT
Steven:	i stand up behind Brian AND rub my dick all over his ass
Jackie:	oh my ass yes yes yes
Gina:	::smiling at Sonia and Beverly::
Sonia:	licking Bev's pussy gently
David:	guiding my cock in from behind
Steven:	and i kneel back down
Larry:	watching Steven...
Sonia:	standing and walking to Gina
Gina:	::licking my lips::
Jackie:	mmmmmmmmm
Gina:	playing with my pussy
Brian:	pumping her cunt wondering if she wants it in the ass
Steven:	spreading Brians legs wide
Jackie:	D push it all the way in
Debbie:	OH YES, PUT IT IN BRIAN
Sonia:	leaning down and kissing her deeply
Steven:	i slide under him and start to suck on his balls
David:	pulling her hips back onto my long shaft
Gina:	oh Sonia
David:	fucking her harder
Jackie:	mmmmm all the way now
Debbie:	AND FINGER MY WET PUSSY

NAME: BiBiLady, **CITY/STATE:** Houston, Texas, **STATS:** Very married very bi, female, **DESCRIPTION:** meeting sexy bi females to join me and hubby in some great times.you wont be diappointed…. Life is too short…enjoy it while u can…the soft kiss of another woman can not be duplicated…not into couples or single guys so pls dont reply…single bi or lez fems or married fems who can party without hubby…all else pls

NAME: SlaveMaker, **CITY/STATE:** Mass, **STATS:** Divorced Bi Domme, 34 years, 5'3" 108 lbs, 34C, **DESCRIPTION:** inshape attractive bitch always looking for new f slaves to follow my commands and do my bidding. Send your gift and we will see if we will talk. Bring it over here honey

NAME: BiGoddess, **CITY/STATE:** Wherever, **STATS:** Female, Sexy, Bi, **DESCRIPTION:** single bi female and luving it…sex, shopping, sex, flirting, sex, men, sex, women, did i mention sex!!! find a penny pick it up, if you're lucky you might get f**ked…heads or tails it doesn't matter!

And, that ends our travels through the female Bi/Gay Chatroom world. Even though most of the profiles are short, they are very much to the point in their preferences, and I am in awe at the number of bi/gay women who want to include a third partner into their relationship. Knowing this type of preference exists in quite a few of the male/female relationships I found it interesting that it was so prevalent in the female/female relationship, also.

The bi female I found to be very open in her sexual preferences compared to the gay female. "Anyone, Anytime" could rightfully be her motto, as she did not seem to be to particular in her quest for fulfillment, as I'm sure you noted in the last chatroom chat.

No matter how you view this lifestyle, it is a preference for millions of women, and one that has also existed since the beginning of time. It is not something new, but a lifestyle that has finally come out into the open along with the lifestyle of the gay man. It never ceases to amaze me how many people think this sexual preference is something "new", and are hoping it is nothing more than a passing fad. Maybe those people should reread the ancient history books they obviously missed in school that discuss this lifestyle, along with all the orgies that were so commonplace in those days. Sex between men, and sex between women is not something new, but something I believe was hidden under a rock when our puritan ancestry came to this country, and something that was never discussed openly until the 20th Century. "Coming out the closet" may be the wrong term for the gay lifestyle. A more appropriate slogan might be: "We have ALWAYS been here, so get over it!"

STR8 SUBMISSIVES

The submissive female chatroom has been a lesson in living a fantasy lifestyle, for me. Here you have a group of females who, either through brainwashing or wanting a fantasy life so intensely, have come to the conclusion that even though they are being controlled and tortured, they are the ones in complete control of the lifestyle. I myself find it extremely hard to believe that when one is tied or chained to an object, having different types of devices attached to their bodies, and made to relinquish their entire being to another human being, that they are the ones in control of the situation. I would surmise that once the ropes are applied, you have lost complete control of the situation no matter what the dominant partner would have you believe or how many "safe" words you use.

These acts are always accepted by both partners under the guise that one could not possible do such punishment, or accept such punishment, if the partners were not completely trusting of one another. You hear the words trust and love uttered like a child begging for more candy in this lifestyle. To the people who live this lifestyle the acceptance of total domination and humiliation seems to be the greatest act of love possible, from the submissive.

These women also seem to be stuck in a time warp, mainly the medieval times, as their profiles and speech will provide for your own judgement. The words "Sir", "My Mistress", "Master", "Maiden", are uttered without hesitation in these chatrooms. It seems to be a pure fantasy, mystical, experience that keeps them living and enjoying this lifestyle. At any moment one would expect a Unicorn or King Arthur to appear. Although I do believe if King Arthur would have tried this with Guinevere she would have dropped him to the ground in the blink of an eye, round table or no round table!

In viewing the different web page links associated with submissive females I have discovered pictures of torture devises, burning of female body parts, actual cutting of skin, and other insane acts all in the name of love and trust. No picture even remotely resembled a beautiful gossamer scene from the land of Camelot.

Even while such torture acts are being performed on these women, they will avow to the end that they are the ones in control and this is the premier expression of their love for the dominant partner. Also, they avow the dominant partner only has their best interest at heart and that they MUST be taught to obey his every command no matter what, for he is only expressing his great love and protection for her.

The majority of these chatroom occupants are either living with their master (knight in shining armor, to be exact), or are looking for a master they can be collared by. As you can see by most of the profiles, these women seem to prefer wind swept clouds, mystical gardens, a knight riding into their life on a magnificent steed, and being rescued from their modern existence to reside in the Never-Never Land of a dungeon. Maybe it is a past life memory from the gothic era that has carried over into this lifetime as a subconscious reason for living this lifestyle. If it is, I would tend to believe the bad memory part has been misplaced by a beautiful fantasy memory.

One statement needs to be emphasized here. **"Please be extremely careful if you enter these rooms."** Not all dominant males that frequent these rooms are knights in shining armor. In fact several men have been arrested for luring women from chatrooms to meeting places, and then raping or murdering them. This is serious business and one should not be stupid while looking for their fantasy adventure in life.

Let us now venture to the world of the Female STR8 Submissive and see her exact thoughts and feelings on this lifestyle.

NAME: BREATHOFSPRING, **CITY/STATE:** All Places, **STATS:** Female, **DESCRIPTION:** She glaces up...that far away look in her eye. She listens carefully for His whisper...like a breath of wind across her wings. She senses His presence...

He is moving ever closer in Her direction. She stiffens...holds still...listening. She yearns to see His face...hear His words...feel His touch. She knows He is coming for Her as surely as She knows the sun follows the moon & stars. She stirs in Her cocoon...wrapped in all the layers of His warmth...still waiting to be set free. She knows that She has always been His...His from the beginning...His till the end of time. He moves ever closer to Her...she struggles within Her cocoon...struggles to be set free. His hand reaches out to Her...She takes it proudly...And so it begins...

NAME: InHisHeart, **CITY/STATE:** In the heart of my master, **STATS:** Female, **DESCRIPTION:** Most of you know me as InHisHeart, He calls me His Little One. I have been captivated by my master, He holds my heart, spirit and soul. Explorer of limits and adventures, I have found what I have searched an eternity for—Him ::warm smile::

NAME: I Bow To Him, **CITY/STATE:** NE PA, **STATS:** Female, Single, **DESCRIPTION:** Preparing to explore and expand my horizons, experience intense sensual passion—to become complete in the eyes of the one to be called "Master". Master and sub, is like the Yin/Yang, both sides being equal, forming the perfect whole. My submission to my Master shall be expressed from "My Soul". Submission goes much deeper than the mere physical/sexual act of offering myself to a Master for pleasure. The pleasure of love, is in loving and serving. I want to be taken over mentally & by the passion and desires he will arouse in me. He will find my inner power which will allow me to offer myself completely & surrender all control to him, on my knees.

NAME: Pleasure Slave, **CITY/STATE:** Northern IL, **STATS:** Female, real time collared slave of my Master, bound by blood 6/19/99. **DESCRIPTION:** 24/7 with Master, by his side in all things, I believe and have faith in the power of US! Making Master go "Mmm", trying to find the invisible binds! Slave, friend, not a push-over by any stretch, "sister" to very few. He who can build the desire is truly the Master! A closed mind is a terrible waste. In this lifestyle that expects tolerance and understanding, there sure are a lot of us running around casting judgment! When you're perfect, take your shot.

You will note that some of the profiles state they are slaves, while others state they are submissive. In reading and talking with submissives, there really doesn't seem to be much of a difference between the two. As hard as I try, I really wish I could believe all of these women, that their lives are this perfect and that the love they write about for their master is really such. But, being the age I am, and having lived through several lifetimes, or so it seems, in relationships, I find it extremely difficult to accept their words at face value. Maybe I'm the cynic here or maybe when you are completely controlled by someone else you really can believe that it is the perfect love you are living.

NAME: Perfect4Him, **CITY/STATE:** By His side where I belong, **STATS:** Female, Collared real time, **DESCRIPTION:** Assuming the PERFECT Butterfly position … Kneeling, legs spread 18 inches, arms clasped behind my neck, back arched straight, eyes lowered in full submission … heart full of love, adoration & lust. ABSOLUTELY NO E-MAIL WITHOUT PERMISSION OR CONSENT…RESPECT MY WISHED & I SHALL RESPECT YOU. His Loyal, obedient submissive … is there anything else in life? He is my life, my love, my world! Master, I will lvoe & adore you forever … my life is in your hands … take me and mold me for your pleasure as I am yours and yours alone...now and forever!

NAME: Innocent Submissive, **CITY/STATE:** NC, **STATS:** Female, 32 years old, single, **DESCRIPTION:** Naturally submissive…at times childlike and innocent, and at others, His mature slave. Yielding to His soothing, exacting voice, I will lovingly submit to Him with reverence, obedience and complete devotion. In exchange for my servitude, He will honor me with love, protection and guidance. He will be the center of my world...my reason for living. Although I crave the kiss of Your whip, Your whip is but an object. It is the working of Your mind that gives me the most sensation.

NAME: IwillObey, **CITY/STATE:** The Heart of the Dragon, **STATS:** Female, **DESCRIPTION:** My heart is protected with Honor. Fear is washed aside and aching becomes pleasure in the liberation of surrender. Opening myself up to him, releasing my fears into his hands, watching and learning as he takes my hand gently turning me in the direction of those fears, showing me how unnecessary they are under his guidance. Where others have tried to take me with demands and loud roars, he was able to with mere whispers—my knees.

Here is our chatroom chat discussion on who is in control in the dominant/submissive lifestyle. Even dominants and submissives aren't quite sure who is in charge. Interesting to know someone agrees with me. <gbg>

A CHATROOM CHAT:

Lit'lLady:	Master4U I am a sub but I wonder if I could be a slave if my situation was different
Master4U:	why would any master accept less than all from a slave
Dom22:	Sinatra's theme song was "All or Nothing At All"
Master4U:	too many ifs Lit'l
QuietOne:	I dunno what I am... <sigh>
Lit'lLady:	yes I know that Master4U and I still wonder
SavyLexa:	And Master4U, I never "forget" or give up on something that I'm bound and determined to achieve
Dom22:	if you need to know, you will find out
Master4U:	that remains to be seen MLady :)
Petal:	Just give yourself time to figure it out Quiet...
SavyLexa:	LOL eat your heart out Master4!
QuietOne:	yes... as the clock spins
Dom22:	the clock can be stopped
SavyLexa:	For you, Big Daddy...I'd come see you alright, as long as you promised to NOT sew me shut.
Master4U:	i am not waiting MLady :) my three keep me busy

80

QuietOne:	indeed
SavyLexa:	at least not my lips, the ones I kiss with. LOL
Dom22:	but your sphincter must be stapled
QuietOne:	acckkk
SavyLexa:	aaahhhhhh!
Master4U:	lol
SavyLexa:	holy sh*t Dom22!
SavyLexa:	LOL
LearningDom:	Dom22 even I a novice Dom know that it's the sub that petitions the Dom
Lit'lLady:	Do any of you Know Master4U??
Dom22:	no...you mean NO sh*t :)
SavyLexa:	uhhhhh, that too
Master4U:	ha
Dom22:	yes...the slave petitons the Master, Learning :)
QuietOne:	petitions = begs...?
LearningDom:	but in the meantime I have no sub where do I go
Lit'lLady:	But the Dom has to accept the Sub or Slave
AngelWings:	<— saw Him on Ameriaca's Most Wanted las week, but other than that...
Dom22:	however...in "mainstream D/s"...
Petal:	lol thanks hon*
Dom22:	the "submissive" is usually the one in control
Dom22:	the Dominant is not always the one who holds the flogger
Dom22:	the real question is...who is in control?
Dom22:	whose IDEA was it?
Dom22:	One commands. The OTHER obeys.
Dom22:	If a "dominant" has to "earn her respect"...guess who is in control? LOL
LearningDom:	Dom22 that something about the power transfer huh
Lit'lLady:	Even being a sub I still believe that the Dom has more control
Dom22:	that is all new age lingo, Learning
Master4U:	sometimes obedience is so seemless that a command is not even necessary
Dom22:	not your fault
Dom22:	it's all over the place
Dom22:	"power exchanges"... "gifts"... "roses and thorns" yuck
Dom22:	it's BS

SavyLexa:	I agree
Master4U:	double yuck
Lady Nancy:	roses and thorns? I haven't heard of that one
Dom22:	yes you have Nancy LOL
Lady Nancy:	lol...I have?
Master4U:	i don't want your gift
Dom22:	the book, yes?
Lady Nancy:	oh
Petal:	gimme da thorns
Lady Nancy:	I have heard that mentioned
Dom22:	wake up and smell the thorns LOL
Lady Nancy:	LOL
Petal:	big book w/ that guy Steed in the pics... lol
SavyLexa:	LOL
Dom22:	btw, the author is good people
Dom22:	but the book is nonsense
Lady Nancy:	I haven't read it, obviously
Cloud Softly:	you really think so Dom22?
Dom22:	historically, it was part of a trend to make "D/s" acceptable to the mainstream
Dom22:	unfortunately, it worked
Lady Nancy:	why should it be acceptable? I mean, who cares?
Dom22:	the concept was that D/s is very similar to "romance" with some kink added
Dom22:	that is what the "power exchange" was about
Dom22:	public relations
Dom22:	eg: "all relationships are "power excahanges"...
Petal:	no chit
SavyLexa:	geez, why do people give up so easily?
Dom22:	just in some, the exchange is a bit "lopsided"
Lady Nancy:	"Power exchange" always reminds me of one of my son's Nintendo games
Master4U:	that is how most are MLady
Dom22:	I want them to "give up" MLady
Dom22:	if they can be convinced to give up from pixels...
SavyLexa:	point taken
Dom22:	how long would they last RT? :)
SavyLexa:	you two are right
Master4U:	but is a nonsensical trem
Petal:	Oooo pixel me baby
Master4U:	power is not exchanged

Dom22:	power exchange= I take, they give
SavyLexa:	from the moment I look at a dom
Dom22:	there's ya "power exchange"
SavyLexa:	who is in control.
Dom22:	and TPE? That's for indians
SavyLexa:	LOL GM
AngelWings:	LOL Sir
Petal:	That term just plain does'nt make sense
Dom22:	well, it does from a nilla perspective
Lady Nancy:	SSC is the one that makes me wanna puke
Dom22:	and that was the idea
Master4U:	no kidding Nancy
Dom22:	the intent was good...
Dom22:	"acceptance"
Dom22:	but, in being "accepted", the original concept was corrupted beyond recognition
Dom22:	"D/s" is the Lonely Hearts Club for the Oughts
Petal:	That's true... but i've just never understood how if it's total it can be an exchange.
Master4U:	of course you do :)
Petal:	unless i've just always misread the term.
SavyLexa:	almost an oxymoron, Petal?
Dom22:	lotta romantics "lookin for love in all the wrong places"
Petal:	mm., exactly MLady
Dom22:	no...you've read it right
Dom22:	it's nonsense
SavyLexa:	I quit going to the club here...
LearningDom:	Dom22 the power exchang is simply a consent and submision
SavyLexa:	the people are fakes. most of them anyways.
Petal:	But since i've never really bought into terminology i didn't let it bother me too much. ;)
Dom22:	the "power exchange" is a new age catch phrase and buzz word
Dom22:	terminology doesn't matter
Dom22:	but the concepts do
Dom22:	I'll remind y'all to go to:
Dom22:	www.XXXXXXXXXXX.com
Master4U:	lol
SavyLexa:	LOL

Petal:	Of course but the focus has gone from the meat to the terms used to describe it.
Dom22:	especially useful for new submissives/slaves
Petal:	oh of course lol
Dom22:	and nice pics for old timers :)
Dom22:	and FUN for the whole family!
Dom22:	wel cum MLady, feel free to mail me after viewing the site
Dom22:	hugs spanks and licks to all what needs em
SavyLexa:	thank you Dom22, I will =)

NAME: Sub2Rule, **CITY/STATE:** Anywhere, **STATS:** Female, single and not interested, **DESCRIPTION:** Love is but an illusion centered in the eyes of the naïve…It is a glimpse of what will never be. Sex is where its at. either control or be controlled. There are many avenues of sex…What is your pleasure? Yeah D/s is not a game, and, so, that established, Mind Sex is the most powerful type of sex I think. Once you have them where you want them you can twist and bend them to your will with a simple glance "nice thought" LOL just the SAM in me. But without Trust you control nothing. Once trust is broken that person will not give themselves to you completely. So beware of treading on someones feelings. It may come back to haunt you. A woman is like a tea bag. You never know who strong she is until she gets into hot water. —Eleanor Roosevelt

NAME: CanSetLimits, **CITY/STATE:** NY, **STATS:** female, mom, single, **DESCRIPTION:** Top/Sadist/Friend…no cyber…no phone…no marrieds…locals only and no, Montana is not local…no exceptions. This is about reality for me, not some far fetched fantasy. Ya wanna be my friend…be real…be honest…be open…know how to laugh. Lifestyle attitudes are a dime a dozen…get over yourself and get a real job. Serve me up some pretty pretty people…serve me up someone I can believe…Not looking for emotional entaglements or a houseboy. Proud Mom 1st, criminal justice…The Art of SM. I have incorporated all my interests into the best of everything…a balance…Actions speak louder than words…Good judgement comes from experience…Experience comes from lots of bad judgments. Yes I'm quite experienced. Pain is involved.

NAME: I Follow, **CITY/STATE:** Anywhere, **STATS:** Female, **DESCRIPTION:** "When love beckons to you follow him, Though his ways are hard and steep. And when his wings enfold you yield to him, though his sword hidden among his pinions may wound you. And when he speaks to you believe in him, though his voice may shatter your dreams as the north wind lays waste the garden. For even as love crowns you so shall he crucify you. Even as he is for your growth so is he for your pruning. Even as he ascends to your height and caresses your tenderest branches that quiver in the sun, so shall he descend to your roots and shake them in their clinging to the earth. Like sheaves of corn he gathers you unto himself. He threshes you to make you naked. He sifts you to free you from your husks. He grinds you to whiteness. He kneads you until you are pliant; and then he assigns you to his sacred fire, that you may become sacred bread for God's sacred feast. All these things shall love to do unto you that you may know the secrets of your heart, and in that knowledge become a fragment of Life's heart."

NAME: SubGirl, **CITY/STATE:** Clearwater, FL, **STATS:** female, **DESCRIPTION:** My submission is very dear to me. Please don't bother asking for it, it is safely in the hands of another. Thank You, Sir. Dominance and submission are two very special gifts. Neither the gift of Dominance nor the gift of submission can be given without the deepest of Love, Trust, Respect, Honesty, & Honor.

NAME: ForHimOnly, **CITY/STATE:** Michigan, **STATS:** Female, **DESCRIPTION:** He spoke now. His strength resounding in that voice. It caressed her. Calmed her. "Now My Lady. All the walls you have placed in front of me, I have torn down, Do not resist me anymore. You belong to me" ... Where he leads I will follow. My Submission is the essence, the all of me. I will enhance his dominance in all aspects, and walk by his side. It is where I belong. Your melody bears a silent tale. Your promises are a silver tongue. Before I buy what's for sale. Then let it be written and let it be done. Do not mail me without permission. Otherwise, I have no interest in what you may say.

Here is girl talk on trying to decide if a submissive should change to please the dom. Or, if changing is too vanilla. Or, why does a dom try to change us. Or, should we let a dom change us. Etc., etc., etc.

A CHATROOM CHAT:

Ldy Alice:	being definite I can't stand someone saying "So... what do you want?"
Miss Angel:	\<snicker\> I can relate ;-)
LordVon:	Diana is that control or self confidence or a combination?
Ldy Anne:	if a man (excuse me, Dommes, but this is my
Ldy Anne:	context) doesn't have self control, how can he control me?
IBowToU:	that's it Diana,,, it's knowing you can let Him handle it,, and you have no wories

Miss Angel:	someone TELLS me i "have" to do something, I'm gonna tell them to take a flyin leap
Miss Angel:	BUT... with the right person... all i need to know is something would please him and it's done
MstrRichard:	but when they make if your idea Angel...
Ldy Anne:	exactly, Angel!
Ldy Alice:	a dom/me who "makes you" (instills) need to please him/her is wonderful but a tough one
Diana:	LordVon?...it self knowledge...and self respect...from that comes the self control...and then my
Diana:	respect
IBowToU:	but isnt' that why we are like this Angel,, because
IBowToU:	we love pleasing whom we respect and love?
Ldy Alice:	so many put it on like gloves for play or let's take out this set of habits
LordVon:	i see and i agree Diana
Ldy Alice:	I went out w/ a dom (Sigh could he ever... well never mind) BUT
LordVon:	lol
Ldy Anne:	Ldy Alice, that's more bottoming, imo... but that's *my* definition
IBowToU:	lol Ldy Alice
Miss Angel:	agreed IBow...IMO, tho, far too many
Miss Angel:	Dominants try to "change" a submissive—
Miss Angel:	make them who they WANT them to be
Ldy Alice:	as wonderful as he was he was like a light switch when we would
Diana:	amen IBow
Ldy Anne:	been there, Ldy Alice... didn't like it!
Princess:	isn't that the idea, Angel? "i am who my Master wants me to be"
IBowToU:	i can't change whom i am,,, i can only be me,,,
Ldy Anne:	Princess... you're still "you"
Diana:	yes Angel...and imo...that's just another play thing...the i'll change him/her
Miss Angel:	Not in my opinion... I'm me—if it's
Miss Angel:	not ME he's fallen for—what's he even with me for?
IBowToU:	too many i know want to change me,,,,few accept me as i am

IBowToU:	<~~wants to be who i am and please my master being me
Princess:	that sounds like a wonderful vanilla relationship-
Ldy Alice:	I am getting a little sad and misty eyed but
Ldy Alice:	one common thing I think now (and more so as I "age")
Ldy Anne:	there are things to be changed and things not to
Ldy Anne:	be changed... the wise Dom knows the difference
Diana:	IBow...same for me...i will change and grow till the day i'm toes up...
Miss Angel:	Actually,... I think it's much more realistic
Ldy Alice:	is "ignorance is bliss" and "I wish I didn't know now what I didn't know then"
Ldy Alice:	cause when you meet someoen you mesh perfectly with all the others
Ldy Alice:	dont' "fit" and I would rather never have found that yet—have it be out there
Ldy Alice:	so that I could be happy as I dont' feel it anymore
Ldy Alice:	no need to submit/please just boredom dissapointment
Princess:	i think there are parts of me that attract him—but that doesn't mean he doesn't want more for me
Miss Angel:	forcing/making someone change to please you, only makes that person bitter in the end
Ldy Anne:	isn't that kinda like the "better to have loved and lost..." thing?
Ldy Alice:	yeah that saying I am not in agreement with
Ldy Anne:	I think that depends on how far removed you are from that situation
Ldy Alice:	very ture Ldy Anne

NAME: Lady of VA, **CITY/STATE:** Richmond, VA, **STATS:** Female, Collared 11/17/94, **DESCRIPTION:** Total submission and slavery to My Southern Master for all time … which deepens daily in ways I never imagined possible...our computers placed me in His hands, helped Him fine me,,,ownme,,,through His Dominance He made me what I was always meant to be,,,His slave...I was looking...He found me...I am complete.

NAME: Forest Dweller, **CITY/STATE:** WI, **STATS:** Female, Taken, **DESCRIPTION:** I am NOT intrested in mail from Dom's looking for a sub...i will Not meet you or entertain you with chat about what I like…A Big Black Worf they call Yellow Eyes resides in the shadows…watching ever closely over me. It will be a magical place…As he leads me into His mystic…I surrender…body and soul…bound...protected.

As you have probably noticed, I haven't had too much to say about these profiles, for I feel they speak for themselves. Most of these women live this lifestyle in real life, but I'm sure some are just chatroom wannabes who only live it on the Internet. Living it on the Internet and living it in real life are two completely different aspects, as one is perfectly safe typing in front of their computer screen. Actually participating in this lifestyle can leave oneself open to all types of injury, not only physical but also mentally. Extreme caution is the word of the day for this lifestyle.

NAME: MastersLady, **CITY/STATE:** Texas, **STATS:** female, married and loving it, **DESCRIPTION:** I do whatever Master has in mind for me. that thing that allowed my Master to find me, that thing he will take away if I don't behave, oh yes I know about that thing!! Kneeling before him (being the lady my Master wants me to be. If you mail me, come at me like the lady I am or not at all!! "The span of my hips…the swing in my waist…the ride of my breasts" are to be appreciated. I am not a stick figure woman and nor do I have a wish to be one okay.

NAME: PossibleLady, **CITY/STATE:** Everywhere, **STATS:** Female, **DESCRIPTION:** I am not some heaving-breasted submissive dying to star in your gothic romance. I believe in natural selection. I have accepted my place as a strong survivor. I know I will love and trust again…it's part of the cycle of life; it's Natural Law. I am not a liar. I am not a brat. I am not intentionally disobedient or disrespectful. I am questioning but loyal. I am passionate and direct but difficult. I crave attention. I am honest. I am flawed. I have been told that I give good headache. Above all, I am his.

NAME: FindMe, **CITY/STATE:** Southern Belle, **STATS:** Female, uncollared, **DESCRIPTION:** Seek my mind to touch my heart…and find the depths of my soul. He will understand me as a person…take pride in me as a professional…respect me as a woman…and cherish me as his lady! He will appreciate my intelligence and enjoy my conficence. (Heed: Being a submissive does not mean I am a masochist or a doormat) I will kneel before Him in complete submission...not out of weakness...but in strength...for H will have earned my total trust. He will be my friend…my lover…my mentor…my Master. I have drifted in darkness without Him. I await his smile. I will no longer fear my future because of my past…For He will be beside me.

"WithHim"…I will be blessed, "For Him"…I will be eternally grateful, "To Him"…I will be ever faithful. Simply said…"He will complete me!"

The below chat has to do with collars. Collars are a sign of belonging to a dominant, be it in the male or female submissive role, and are actual collars that are placed around the submissives neck by the dominant partner. To be collared seems to be the ultimate goal in this lifestyle and one most submissives are striving for. Similar to receiving a diamond and platinum wedding band.

A CHATROOM CHAT:

DOMRESPECT:	what about creepy wowmen who are too cheap to even buy their own personal toys?
BelongsToHim:	personal toys like BOB?
DOMRESPECT:	and collars, floggers...etc...
Judy:	<—will pass on the communal personal toys. Ewwww
Judy:	hey, i thought you guys were supposed to buy the collars!
DOMRESPECT:	work collars
DOMRESPECT:	silly
GentleCat:	Personal toys are just that PERSONAL
MandyH:	work collars?
Judy:	sheesh. creepy men who are too cheap to buy toys
PrettyLady:	Whats a work collar?
BelongsToHim:	and the womn who love them?
Lord4Lady:	Sounds like a movie, Pretty.
PrettyLady:	I never had a seperate collar—is that unusual?
BelongsToHim:	I never heard of a work collar
Woman in Room:	i have a work anklet
Judy:	im not buying myself a collar. that seems way desperate to me
DOMRESPECT:	collars used during bdsm play
Woman in Room:	does that count?
DOMRESPECT:	dress collars, work collars, everyday collars
BelongsToHim:	I agree Judy
Judy:	if some guy wants to restrain me, let HIM figure out the logistics
PrettyLady:	I've never bought myself a collar—was given
PrettyLady:	to me by Him.

DOMRESPECT:	puurchasing a work collar for a good fit isn't a logistical issue
Woman in Room:	Judy... hell yeah... if he can't afford dental floss... he can't afford you!!!!
Lord4Lady:	LOL Judy - he should have been a Boy Scout
Woman in Room:	i feel your pain!!
Judy:	damn right, Lord. plus they know all those good knots :-)
NightMagic:	<— former cubmaster <ewg>
Judy:	i dont need a "work" collar
DOMRESPECT:	I bet ya didn't even buy any personal floggers, crops, canes
Woman in Room:	Jeff got kicked out of boyscouts... he was running a bar on a camping trip
CHARM:	anyone want to fuck
Judy:	i hear your mom does
BelongsToHim:	Charm...now that is real nice
Lord4Lady:	Really, Woman?
PrettyLady:	Sorry my computer doesn't fit up there.
CHARM:	she's dead
Woman in Room:	damn Charm... get straight to the point
Judy:	then shes probably not real good at it by now
Woman in Room:	Lord... lol... really... true story
Judy:	why do i need to buy toys? i havent even played in a year
CHARM:	judy I bet your good at sucking cocks
Judy:	yes, as a matter of fact, i am
NightMagic:	who else thinks Charm is male?
MasterRyan:	I'll let you know if she is Charm
BelongsToHim:	who cares
MasterRyan:	<smirk>
Woman in Room:	Judy... i got lots of games... we'll play and play
Judy:	i didnt say i couldnt play. i said i havent
Lord4Lady:	Do you ever travel east, Judy?
MasterRyan:	hmm... self abstaining masochist... interesting kink
Woman in Room:	shoots and ladders... checkers... cards... cowboy and indian
Woman in Room:	you pick
DOMRESPECT:	a year?...sheesh
Judy:	umm, not in a few years, Lord4Lady

DOMRESPECT:	anyone who swats ya is gonna get a face full of dust!
MasterRyan:	kinda makes your floggin arm twitch a bit, dont it?
Judy:	heheh

NAME: Single Lady, **CITY/STATE:** Everywhere, **STATS:** Female, Single, **DESCRIPTION:** "Nothing is so strong as gentleness, and nothing is so gentle as real strength." Not interested in married men in any state or country. When we surrender, we don't surrender to a man. We surrender to a part of ourselves that is softer, less controlling, more interested in peace than in argument. It's not a game we play; it's a dance we do. Surrender isn't a losing position in bed, and it's actually not a losing position outside of bed, either. But we can see this only when we're clear that surrender doesn't mean giving up or giving in. surrender is not loss. To surrender is not to capitulate. To surrender basically just means to relax and let someone else have his or her own strength.

NAME: ToPleaseHim, **CITY/STATE:** Arizona, **STATS:** female, available, **DESCRIPTION:** The collar is sacred. Submission is who I "am", not what I "do". I was created for D/s. I have spent my lifetime grooming and preparing to serve the One I will call "Master". Learning what it means to honor, please, mirror and serve Him. Hearing "I am pleased" is one of the single greatest joys in my life. To offer pure surrender to Him will be my reason for being. Continuation of preparing towards Total Power Exchange…grooming to make Him PROUD to call me His Own. I proudly wear the tattoo of a Fire Flame, embraced with a nameless collar. "None before Thee, master of mine….not even I."

The following profile touched me deeply, as this submissive young lady had dedicated her profile name (her real profile name) to her dominator, who had passed away. She spoke from her soul and made it easy for readers to feel her loss, not only in her heart, but also in her life.

NAME: ToUIKneel, **CITY/STATE:** NY, **STATS:** Female, **DESCRIPTION:** Through the darkness she heard him call her name. "My Precious Angel, come to me. Kneel before me and let me guide out of the darkness and into the light of who you are. Allow me to guide you. To teach you. To love you." "I shall not hurt you My Precious Angel." She knelt before him eyes lowered with gentle tears flowing down her cheeks. In a voice so quiet that only he could hear her say, "Yes Master I am you precious Angel. Today, tomorrow, always."

Donna Tracy

NAME: Lady Jody, **CITY/STATE:** St. Petersburg, **STATS:** Female, Married, **DESCRIPTION:** Owned by the one who has made the word Master more than a noun…He has made it a verb & an adjective. He knew what it took to release me and once my spirit was free, how to conspire with it, corrupting me with my own appetites, dizzying me with my own senses, convincing me that I am so light, graceful and powerful that I can do anything—even fly. Once I trusted Him, He took me to the edge of an imaginary cliff, closed my eyes, and slowly, carefully, turned me inside out, emptying every fear and thought I had, until there was nothing left of me but breath, heat and a core of undiluted self. Then at the first quiver of my muscles, He spread my wings, kissed the warmest dampest feathers and flung me over the side into my dreams, down canyon walls, thru clean blue air, in one exquisite concentrated moment of self love. Master, you have Mastered me.

NAME: ToServe, **CITY/STATE:** Somewhere, **STATS:** Female, **DESCRIPTION:** You know my name, it is "Desire", the feeling you get when you see your lover out of the corner of your eye as she brushes her hair. I am a woman, child, mother, and sister. The truth I hold most precious, is that I have been born to serve men. I am built to satisfy men, physically and emotionally. You are the center of my world, the altar at which I worship. I wait for you to take me to the place where slaves are safe, at your feet, in your bed, and abiding in your every breath. This is who I am, my life was created to serve you, and serve you well. I enter your chamber on my knees and whisper "Master may this worthy slave enter so I may learn from You and serve You." I kneel with head held high, eyes lowered, knees spread for you inspection, and my hands crossed behind my back, in symbolic bondage. The God Eros has blessed me with libido, brains and the desire to inspire you to your greatest. I am a "slave", not a submissive. I have only your will, but will serve only one. My heart is encased in spider's web, as strong as steel as delicate as glass.

Ahhhhhh, I want so hard to believe the words these women have written. They are beautiful. Romantic. Right out of a paperback novel with the painting on the cover of that incredibly handsome man with the long, windblown hair, rippling muscles, and piercing eyes, who is clutching his lady with the heaving breasts to his chest while his steed stands in wait to take them to Happily-Ever-After-Land. Such incredible, beautiful profiles have been written here, but alas, a few things have been omitted or overlooked in these profiles. No one has mentioned the rope burns, the cane marks to the buttocks, the body piercing, the clamps to breasts and other body parts, the carving on skin, and the rituals of degradation and humiliation. These acts are just the tip of the iceberg in this lifestyle. If anyone tries to tell you it is all about love, you had better ask to see the body marks or scars left by this love before you believe them.

If I sound a little hard on this lifestyle I guess it's because I have a real problem with dishonesty. And, this lifestyle, to me, is about the most dishonest of lifestyles out there. I personally cannot equate love, caring, and kindness, with beatings, burnings, and being hung from the ceiling for hours. Somehow the ying/yang of this lifestyle just doesn't fit together in my mind, and I honestly hope that whatever these women are missing in their lives, they find before it is too late. I believe every woman has always wanted, if she hasn't already found him, a knight in shining armor to rescue her and carry her off to his castle. It probably relates back to our childhood when we read fairy tales or saw the movie "Pretty Woman" when we got older, somehow. Just be sure that when the armor comes off the knight that you are not left with a man wearing a suit of leather and brandishing a whip.

STR8 DOMINATORS

Black leather. Stiletto heels. Chains. Whips. Handcuffs. These are the daily tools used by the dominating woman, or dominatrix, to practice her craft. The image the professional dominatrix portrays in a chatroom is one of a cool woman in complete control—flourishing an iron fist, or flogger, at the first sign of disobedience. The standard photos of the dominatrix on web links show a woman who walks around in leather outfits three sizes to small, four-inch heels, and brandishing a whip. She portrays a real woman who is not afraid to crack that whip at the first sign of disrespect.

While some dominatrix in the chatrooms only prefer computer or phone domination, there are those who have their own offices, referred to as "dungeons", and make one terrific living in real life by inflicting pain and torture on the male body. This is not a one-sided office visit, as women pay for this service also. The dominatrix has a clientele who worships the ground she walks on (and sometimes her feet with a little licking thrown in). The cost of such worship can range from $200-$500 an hour. The men, mainly white-collar workers, who frequent these facilities, allow some of the most outrageous tortures to be performed on their bodies and minds. They can be blindfolded, chained to tables, have clamps applied to them, and their buttocks beaten with a cane until they turn a crimson shade. They can be dressed in binding leather, women's corsets, or tied spread eagle while hot wax is applied to very sensitive parts of their torso. They can be verbally humiliated by the dominatrix who will expound on how the man is nothing, and made to crawl at her feet. For this treatment the dominatrix is rewarded handsomely, via cash, plastic, or check.

These acts are usually of a sexual nature for the submissive, with culmination leading to the man being allowed to climax when the dominatrix says he may— which may be thirty minutes to an hour down the road.

The dominant female who lives this lifestyle 24/7 usually shares this lifestyle with a husband who has accepted her collar. It is basically the same living arrangement as the dominant male/submissive female, except for the roles being reversed. Here the female is in control and the male does her bidding. She is treated like a queen and her every wish is granted by an adoring, submissive male. Everything from the cooking and cleaning, to allowing her to chain, whip, or humiliate him, all in the name of love. Some submissives are not allowed to go to bed, use the bathroom, or get dressed without first asking permission. A dominant woman controls their every move.

Many of these females do not let their friends or families know about this lifestyle they are living. They prefer to keep it hidden from the world, not wanting to risk the judgement call of their peers in the workplace, or the backlash of the family. The men in these relationships are treated basically the same way as the paying customer who goes to a professional dominatrix, with the exception that the punishment/love is received from his full time partner or wife. Once again we have a relationship based on pain and torture that is being glorified as proving complete love and trust for one another.

Through the chatrooms I have learned that there are clubs in major cities across the country where dominant/submissive people, or the wannabes and lookers, can go for an evening out at the local dungeon. These clubs are strictly controlled, allowing no sex, be it oral or penis penetration (masturbation is ok), and anyone caught prostituting is quickly escorted out. Most of the clubs are "bring your own toys", or you may purchase toys onsite or from a store that is usually conveniently located just a few steps away and will stay open until 3:00a.m.to accommodate the late night shopper. (See what you can learn by following links and clicking that mouse!) I have included a web link in the appendix, which lists such clubs in many of the states for your use if you so desire.

This lifestyle also has weekly or monthly get-togethers called "munchies". A munchie is a social dinner where dominants and submissives get together for a social evening. Most of these people travel in the same crowd, preferring the safety of friends in the same lifestyle.

I would venture a guess that roughly 70% of the screen names in these rooms had links attached to them. These links will take you immediately to graphic web sites, which there was no charge for. Of course there was a charge if one wanted to join a web site, but the majority were free. The only control that was offered for these web sites was by asking if the viewer was 18. Answer yes, and you are immediately confronted with sexually graphic BDSM photos.

Below are some profiles of dominant women who frequent the dominant female/submissive male chatrooms. Some are computer/phone dominants, some are 24/7 dominants, and some are very professional dominatrix with their own offices. I found these women not too interesting, as a lot of their conversation was of a nature, which seemed to me, to be trying to convince everyone in the chatroom how in control they were and how unaffected they were by mundane conversation. I believe a little "too" much of trying to be aloof and controlling goes a long way.

95

NAME: MistressAlana, **CITY/STATE:** Southern California, **STATS:** Female, married to a wonderful f/t r/t sub., **DESCRIPTION:** READ THIS PROFILE VERY CREFULLY!! Ask permission to IC in the room I am in, but be prepared to accept NO as an answer... if I'm not in a room, don't bother me!!!...If you're married, GET LOST!!! NO one under 40 will be considered!!! NO PROFILE, NO CHAT!! All unsolicited MAILS will be ignored! I'm 5'4", 115 lbs, butt-length red hair, blue eyes, (???)-24-35. If you believe that, I have ocean-front property in AZ for you. You have read my profile...if you cannot follow my rules, you risk my wrath and my flogger..."I pity th epoor soul who relies solely on cyber for their sexual entertainment."

NAME: ForYourPain, **CITY/STATE:** Atlanta, **STATS:** Female, divorced, over 40, **DESCRIPTION:** Sensual Domme...CBT, Floggers,,Crops...Etc. I have a good occupation and so should you
"I don't think you know what pain is...I don't think you've gone that way"

NAME: ControlDom, **CITY/STATE:** NYC, Manhattan, NY, USA, **STATS:** Female, divorced, and happy, **DESCRIPTION:** If you want to apply, be sure to Real Alll of MY profile! I create beauty, spirituality, peace and tranquility. I am controlling, elegant, refined, slender, educated, beautiful and very strong. D/s is the yin and the yang in MY life. NO MARRIED/INVOLVED men. At Least 40yrs of age, NOT YOUNGER, very smart with a great sense of humor, LOCAL, and VERY Handsome! NO Wimps! I am fed up with assh*les who are not subs and/or who think that this is a way to get a quick fu*k. NO ONE tells ME what to do! Send ME email with a close up, G-rated picture and your biography if you meet ALL the above initial requirements be willing to relocate. I AM NOT INTO BE or GAMES, SO, IF YOU ARE, DON'T WASTE MY TIME! GO AWAY & GET A LIFE!!

NAME: MistressEm, **CITY/STATE:** NY, **STATS:** Female, have collared my sub RT, **DESCRIPTION:** Please do not IC me for that Honor I give to very few. Spend time training my puppy dog, training him to learn how pain turns into pleasure and how everything is happiness if looked at right way, of course my own... NO EMAILS WITHOUT PERMISSION D/s reading books and taking more knowledge into my relationship and don't ask if im really into...i have Mistress & pain & pleasure with whip & handcuffs tattooed on me...

NAME: M'Lady Wic, **CITY/STATE:** Louisiana gulf, **STATS:** Female, proud owner of Imperial Sub, **DESCRIPTION:** I DO NOT CYBER DO NOT ASK. Obedience honesty trust friendship from the beginning to the end. My own will rule with tender hands and loving heart but watch for the flogger. Email only and ask for interview or to chat. If married or have problems with bbws move on don't waste my time. I want to play with ur mind not just ur body make u see things never dreamed of before and making u want more I have played in the mind and redden the body do u want to feel it too!

NAME: MistressAdept, **CITY/STATE:** New York, Philadelphia, Miami, Atlanta, Boston, Baltimore, Columbus, Cleveland, Minneapolis, Washington DC, **STATS:** Female, single, **DESCRIPTION:** If you are worthy, I'll show you! bdsm, bd/sm, bondage, domination, dominatrix, erotic punishment, spanking, submissives, fetish.

Even the controlling dominant woman can find dissention in a chatroom. I'm not sure what all this was about, and I don't think the room occupants know either.

A CHATROOM CHAT:

Sweet Dom:	why do people have to use bold and caps?
Freddy:	Will ya dress me in your panties Ms Dom?
DontTalk:	They're legally blind?
DomBitch:	CAUSE WE WANT TO
Ask First:	feed their egos Sweet
DomBitch:	AND IT'S OUR COMPUTER
Ask First:	obviously their egos are starving
DontTalk:	<kicking Dom bitch, aka pissqueen>
Sub Alfred:	they have a geat need for attention
DomBitch:	AND I SEE WE ARE GETTING FROM SOMEONE
DomBitch:	I LIKE ATTENTION
DontTalk:	Like this, Bitch <CLICK>
Sweet Dom:	well, it annoys me and you all know I don't like
Sweet Dom:	to be annoyed
DomBitch:	I HOPE YALL LIKE THIS TOOOOOOOOO
MichaelXXX:	MsSuzin, you are so right.
Sub Alfred:	yes ma'am
DomBitch:	SOUTHSIDE IN THE HOUSEEEEEEEEEEEEEEEEEEEEEEEEEEEEE EEEEEEEEEEEEEEEEEEEEEEEEEEEEEEEEEE EEEEEEEEEEEE
Sweet Dom:	yanno what Bitch, you give Bitches a ad name—so
Sweet Dom:	you're going in the trash heap
DomBitch:	FLA REPRESENTING THE 2000

Ask First:	oh gawd...another wannabee ghetto thug...sheeeeeeesh when will KIDS grow up??
DomBitch:	IT'S MRS. BITCH
DomBitch:	WHEN THE WORLD END
Ms Suzin:	DomBitch earned perm ignore for me too, Sweet.
DomBitch:	WE WILL GROW UP
Sweet Dom:	Michael, I don't ever have to be nice so don't tell
Ask First:	yes m'Lady??
Sweet Dom:	me that
Ask First:	i'm gonna get some COOKIES???????????
MzSouthrn:	LOL Ask First
DomBitch:	FU#^ YALL UP IN HERE
Mistress Peg:	yeah...I got your cookies...LOL
David4513:	Who the truck was that?
DomBitch:	THAT GOT THEY EYE'S ON ME
Ask First:	COOOOOOOOKIES :)
Sweet Dom:	and that other annoyance too
DomBitch:	HI ROOM
No Lady:	searching for my whip? anyone seen it?
DomBitch:	CAN YALL SEE ME
Sweet Dom:	look up your ass NoLady
DomBitch:	I SEE SOMEONE LIKE ME UP IN HERE
No Lady:	Hey Leather, just because your ass is full of gerbals.
Ms Suzin:	Wow, I have never managed to lose a whip
Ms Suzin:	online, how incompetent can you be to lose a
Ms Suzin:	whip in cyber?
Ask First:	oh yeah yes m'Lady make you some cookies
DomBitch:	MAN GO ON A COOKIE ROOM IF YOU WANT TO CHAT ABOUT COOKIES
PrettyLeather:	I THINK IM GONNA TRY SOMETHING DIFFERENT
Mistress Peg:	hey bitch...shuddup
Sweet Dom:	and yet another piece of trash for the heap
PrettyLeather:	AND CLICK No Lady

NAME: Domme2Good, **CITY/STATE:** Detroit, MI, **STATS:** Female, Single looking for 24/7 slaves, **DESCRIPTION:** Cane, leather slapper, gs, cbt, subs, slaves. I'm tired of online game playing liars! So save it. Only the strong survive this!!!

NAME: FeelsGood, **CITY/STATE:** US, **STATS:** Female domme, **DESCRIPTION:** Disciplining and beating worthless slaves, all disciplines of BD SM and kink. DOMINATRIX. DO IT NOW SLAVE!

Boy these women are tough! I wonder if that is really true in real life or just here on the computer where it is safe to be tough. Even though women's lib came in years ago, I don't know many women who go around this tough on the outside. Could all this toughness be one great big bluff?

NAME: PleasureGiver, **CITY/STATE:** All four seasons, **STATS:** Female Mistress, **DESCRIPTION:** No instant chat without permission!! If I am in a room ask, if not send an email, if you can't follow this simple request then I have no interest in talking to you anyways! YOU WILL BE IGNORED! CBT, Wax, Cropping, Flogging & Bondage are what I do. I keep it Safe, Sane & consensual! Single or divorced men ONLY! Between the ages of 30 & 50. Do Not Waste My time on excuses or on how your wife/girlfriend just doesn't understand your needs. I really don't care! NO sissys, mommy boys, whiners or brats! Some call it Kinky or Perverted, I call it Passion! Only those not faint of heart need apply. Be witty & interesting and maybe you will hold my attention, the submissive mean that can do that will capture this DommesY

NAME: SubTrainer, **CITY/STATE:** Anywhere, **STATS:** Female Mistress, **DESCRIPTION:** You may call me Mistress, respectfully. If you would like The Mistress to consider an appointment with you, e-mail me. I reside in the land of dominance disipline control and teaching you obedience and respect. There are many avenues in my neighborhood which we can explore together if you need a strong FemDomme to lead the way. I live in my world and may consider seeing you. The Mistress needs no Partner!!! Your Disipline, Your Pleasure. spankings Bondage Foot and Leg Fettish (mine of course) Forced Feminisim (You BAD girl) Male and Female submissives considered for private sessions. No sex involved. I am The Mistress I am not your sex toy. I will only consider seeing you on my terms and at my convience. A 1 or 2 sentence e-mail will not generate a response from me. I am a Professional Dominatrix. If your e-mail interests me I will respond and send the information you will need to fulfill your dream or your fetish. All sessions are safe and strictly one on one unless otherwise pre-arranged, to assure your privacy and mine.

NAME: Dominatrix OF Men, **CITY/STATE:** here/there, **STATS:** female Mistress, **DESCRIPTION:** Various torture devises and methods. I am a Dominatrix and your pain is my pleasure!!

NAME: Mistress Pamela, **CITY/STATE:** The dungeon, **STATS:** Female, single—I never get tied up!, **DESCRIPTION:** Dominatrix. Torturing men with pleasure…kneel before your goddess and worship me!!!!

NAME: Pro Specialist, **CITY/STATE:** London, UK, **STATS:** Female, Dominatrix, **DESCRIPTION:** I am currently accepting humble applications from genuine submissives for **real-time** sessions in my London chambers www.XXXXX.co.uk. I currently have several slaves in my possession. I am a Professional Dominatrix and Behaviour Modification Specialist; Beautiful Bitchgoddess. "Bow down before the one you serve, you're going to get what you deserve."

The next profile is just one of thousands who do their advertising on the Internet. In every type of lifestyle preference, sex is for sale on the Internet. But, from what I have come across in my travels on the Internet, why pay when you can get it for free?

NAME: Kneel4Me, **CITY/STATE:** Atlanta, GA, **STATS:** Female, 5'8", 150, Auburn, Green, Scottish Irish, fair, MBA, SMU, single, **DESCRIPTION:** Bright, accomplished, men with special submissive needs. They must be available, over 40 and single if local. The price is equal to the reward. That place you are when you know you have no choice. You must yield. You must please. YOU ARE MINE. The key to the power exchange is having some to give me. Guiding submissive men on a journey through pain to pleasure. I take power and turn it into exquisite pleasure…. MINE! I know what I want and get it, at any cost… Yours most often, with a SMILE! Acceptance, nurturing and knowledge are the keys. Enter my web at your own risk. www.XXXXX.com There is no turning back. The dominatrix in me has done her job well.

NAME: TeacherDom, **CITY/STATE:** Cleveland, Ohio, **STATS:** Female, happily involved, Mistress, **DESCRIPTION:** Erotic Power Exchange, Submissives, Behaviour Modification, Crossdressers, Bondage, Humiliation, Sensory Depravation. Tribute to Mistress is a must. Presents are nice….know your place, if you wish to speak to me you may email me only! Professional Dominatrix—do Not waste my tiem if you are not serious!!! If you wish to speak to me you may email me. "The Good Life is waiting for us here and now!….at this very moment we have the necessary techniques, both material and psychological to create a full and satisfying life for everyone." B.F. Skinner

NAME: PhoneDom, **CITY/STATE:** WA State, **STATS:** Female Dom, **DESCRIPTION:** My web site is available for those who are of 21 or older. In the realm of fantasy there is only one place you seek to be…finding your desires in Mine. I own many subs…r u next? Teaching subs the pleasures of Obedience. If you are not sub or serious move along, I am seeking those who desire to worship Me from head to heels. Silky voice w/leather attitude, I am a devilish Dominatrix with 7 yrs exp, 2 yrs phone. Your bound to love it. Are u ready to follow my commands? Domina by Phone, ask for site info by e-mail/with hello and age for appt. 21 and up, PHONE ONLY. NO R/T, CYBER or IDLE CHIT CHAT! "I would never harm you, but there is a world of difference between hurt and harm My dear" M.C.

NAME: Spicy Dom, **CITY/STATE:** PA, **STATS:** Female, Single, **DESCRIPTION:** Erotic and hardcore encounters, the pleasure of leather, teaching big boys to behave like little ones. Thigh-high leather boots, spike heels, leather and PVC corsets, lace thongs, and always a miniskirt and nylons when in public. Professional Dominatrix (animal trainer of sorts). "Never met a man that I couldn't tame, train, or restrain" "You look so cute in those ruffled sissy pants"

NAME: DomWithAttitude, **CITY/STATE:** Everywhere, **STATS:** Female, forever free, **DESCRIPTION:** Enslaving men, sex, dominatrix, mental mind f*cking, oral c*pulation. Part time porn star, full time pain inflicter. "If you have a friend whos kind and true, f*ck him before he f*cks you"

The following chatroom chat absolutely fascinated me. I have waxed my legs to get rid of hair, but never have I had the inclination to apply it anywhere else on my body! Wax is hot and burns—but, maybe that is the idea? At any rate, I think I'll pass.

A CHATROOM CHAT:

LovelyDom:	before you start on that subbie
LovelyDom:	rub him/her down with mineral oil
LovelyDom:	slippery baby
LovelyDom:	Mmmm
LovelyDom:	now you can ladle that wax on any body part
LovelyDom:	we do clit molds
LovelyDom:	cock molds
Good Sub:	<— is entranced with this!
LovelyDom:	all kinds of fun places
LadyXXOO	mmmmmmmmmmmm
Mistress Dora:	:::raising my hand::::: are we talking on bare skin Lace
LadyXXOO	lol Good, Imma tellin Susan

Dominant Luv:	so am i Good
LovelyDom:	it will feel like the warm wax you get at the really good European spas
LovelyDom:	yes, bare mineral oiled skin hehre
Good Sub:	Please do, LadyXXX!
Good Sub:	lol
LovelyDom:	do not worry about body hair
I Rule:	LovelyDom, I was about to say...it's just easier to buy the religious candles and...
LovelyDom:	you can do this anywhere LadyXXOO mmmmmm to hell with the subbie Lovely...I'm gonna use the warm wax on me!
LadyXXOO	rofl
I Rule:	line em up in hot water in big pot
LadyXXOO	sounds absolutely delicious
Good Sub:	Doesn't it though, Ma'am?
LovelyDom:	this is so good...I had 3 fem subbies wax me and give me a spa treatment
LadyXXOO	sure does Good sub
LovelyDom:	let the heat soak in
LovelyDom:	it feels so good
LovelyDom:	and if you have aches... Mmmm
LovelyDom:	now... when you are ready for removal
LovelyDom:	this will all rub and lift off in a sheet
LovelyDom:	to remove any tiny bits that are left
LovelyDom:	shake baby powder over the area
LovelyDom:	then rub briskly
LovelyDom:	then towel the person off
LovelyDom:	they are refreshed, renewed... and it is a wonderful introductory to waxing
Mistress Dora:	I saw a Domme doing this to her sub at a play party (women
Mistress Dora:	only)... the sub went into "subspace" she was so high from
LovelyDom:	that anyone will love
Mistress Dora:	the sensation
I Rule:	gotta buy some good throwing clay and and mix to creamy... fun afterwards to remove oil
LovelyDom:	you should have seen Robert
LovelyDom:	we did his cock, balls, butt...
LovelyDom:	he was completely coated from toes to neck
LovelyDom:	layer upon layer

WillObey:	Hello everyone. 33 sub male WI here.
LovelyDom:	the heat lingers and soaks in
LovelyDom:	I have paid more than $100 to have this done to me at spas and I have to tell you, it is won
LovelyDom:	wonderful
LadyXXOO	<~going to Lovely's for wax treatment...lol
LovelyDom:	doing it within a bdsm setting is really a treat for any subbie
LovelyDom:	LadyXX, I let Robert help me and we did some of our domme friends as a treat for them
LIKE, YA:	do any of you like to step on men with high heels LadyXXOO LOL I'm going shopping for wax and a crock pot!
Mistress Dora:	yeah... I know what you mean... the Domme was kind
ComeHere:	i do
Mistress Dora:	enough to make those watching aware of the fact that her
Mistress Dora:	subbie was not on any kind of drug... just the endorphins
Mistress Dora:	from "subspace"... she peeled the wax off in one sheet and
Mistress Dora:	then used a straight razor on the subbie's already shaven
Mistress Dora:	cunt
LovelyDom:	Mmmm Dora
ComeHere:	boring
LovelyDom:	I still have a clit mold from a fem subbie friend that we did
LovelyDom:	she was in heaven
LovelyDom:	ComeHere!!!
LovelyDom:	I will use beeswax on you, bratty one
LovelyDom:	hmpt
ComeHere:	ok
LovelyDom:	talk about hot hot hot
ComeHere:	make my day
Mistress Dora:	who?
Good Sub:	Well, subbie needs to go keep an eye on the
LovelyDom:	I can tolerate beeswax on the bottom of my feet but no where else
Good Sub:	little domme for
Good Sub:	a bit.

103

Mistress Dora:	oh... the beeswax
Mistress Dora:	duh
LovelyDom:	lol, GoodSub... have fun

This next profile should be inundated with responses from geeks. What with the computer age upon us, just about everyone is a geek these days...at least at the office. Good luck to all the applicants. May the best geek win!

NAME: DomLvsGeeks, **CITY/STATE:** Portland, OR, **STATS:** Female, not a pro, **DESCRIPTION:** if you are over 18 see my website @ www.XXXX.com I am NOT looking to travel nor for YOU to relocate. I have a technogeekfetish which means I have a thing for Csi, Ees, physics, chem types, etc. to torture. Extra points for glasses, pocket pen protector and math brain. I so love to take boys out of their geek brain & into their tormented body. I'm NOT A PRO. I love: Sadistic Torturing, CBT, Flogging, Electricity, Strap-on, Play Piercing, NT, NO CYBER NO PHONE (Yuck, p'tew). I am looking for a full body masochist to scream for me on a regular basis (in person). No faux masochists please! So if you profile as a submissive, masochist geekboy in the Pacific NW (Oregon) area, I want to hear from you. If you meet the criteria AND are emotionally durable, have real-life experience or are willing to face the (likely overwhelming) consequences, you may email me. You need to be willing to suffer absolute agony & the resulting expiation. For all you visual boys, I have a pic but it doesn't mean you get to see it (wanker boys are so visual). If you pass muster, pix may be exchanged.

NAME: Elegant Dom, **CITY/STATE:** Bay Area—CA, **STATS:** Female Domme w/Fetish Family, **DESCRIPTION:** Creating the atmosphere of elegant Dominance and submission and using My creative Sadistic Nature—attending to and Fulfilling Client Fantasies. DO NOT SEND IC without having first sent an E-MAIL inquiring if you have permission. State your intent = Client or Personal. I am a Professional Dominant—Sadist. Reach deep inside and define your personal worth... be able to state what your attributes are—and hold true to your personal integrity without reservation.

NAME: Dom2BeKind, **CITY/STATE:** Greensboro, NC, **STATS:** Female, single, **DESCRIPTION:** Hot wax, floggers, fetish wear, liquid latex, spanking, humilitation, CBT. Real time and local ONLY. Some sweet little boy, do you think you could be him? Tell me WHY. I will make you beg for my dominatin (Im' NOT BBW) BDSM D/x. I am only interested in talking to single, attractive male subs 24-25. Email with pic. Be stable and secure.

NAME: Mistress Ann, **CITY/STATE:** NYC, **STATS:** Female, single, **DESCRIPTION:** Born to diminate, to tease, to please at will… bringing you to your knees to beg for more. Adore leather, silk, velvet & skin. The softest touch with a will of steel … erotic, sensuous, imaginative. The body responds to what the soul desires. Test your limits … explore your emotions. Be prepared to prove your devotion. Goddess … of all you crave, of all you need. In privacy, there is no shame in humiliation, no shame in serving... of accepting & embracing the fate that your actions have brought you.

This next profile causes me a bit of confusion. First the lady states only intelligent, mature men need apply. Then she goes on to say that she will be a mommy to a little boy, with diapering included. Somehow I lost the connection between maturity and little boy.

NAME: MomDom, **CITY/STATE:** New York city, **STATS:** Female, Single BBW, **DESCRIPTION:** I do not play games... married or involved men need not apply. If you do not possess intelligence or maturity—do not contact me. I am not a professional dominatrix. However, I am a highly educated, passionately sensual, honest and demanding woman. Your mommy wants you to worship her, OTK, diapering... all the good things you want with your mommy. R/t only... no cyber. Making sure that my sweet little boy is taking care of his mommy. Sugar and Spice and Everything Nice. Good boys go to heaven and bad boys go everywhere.

NAME: Mistress4Money, **CITY/STATE:** Southern NH, **STATS:** Female, **DESCRIPTION:** In Search Of a Real Time sub male. Applicants must apply (email) with a detailed description explaining how they feel they can serve Me. (if this is too much of a chore for you then you would not be able to serve Me). I'm simply a Princess... a real one. And in my future I hope to be...looking down at not-so-free ones...princes/"boy toys" dedicated to serving royalty. (And I mean "serving" not "getting your rocks off"... there's a difference.) I do not have sexual intercourse with slaves. You must have the need...and the means...to provide financially coupled with a desire to be owned. I seek a "slave-partner" nto a boyfriend...someone who knows the difference between D/s and BDSM.

NAME: LDYDOM, **CITY/STATE:** NJ, **STATS:** Female, 6' tall, BBW sensual, dominant black woman, **DESCRIPTION:** I am NOT LOOKING, so you IC will most likely be ignored. I don't mind chatting with courteous people who are in my geographical area. D/s, some whipping, some bondage …lil a this … lil a that …Seriously committed to the FEM Philosophy (Females Enslaving Males) …FYI—NO CYBER, NO PHONE, NO PROFILE and/or MARRIED = NO CHAT. "Ask not what your Domme can do for you, but what you can do for your Domme" <g> "peel me a grape ... crush me some ice ... skim me a peach, save the fuzz for my pillow ..."

NAME: TX Mistress, **CITY/STATE:** Dallas/Ft. Worth, TX, **STATS:** Female, Feminine, loving, sensual, and sadistic non-pro BBW Domme, **DESCRIPTION:** IC'S BY PERMISSION ONLY. NO CYBER, NO PHONE, & NO MARRIED MEN!!! Please Note: Only interested in those who are intelligent, loyal, honest, 40+, not married, and seeking an ongoing relationship of regular service. Local or relocatable only. INTJ & Enneagram Type 3 Bdsm; sensual dominance; masochistic males; administering a mixture of pleasure and pain, tenderness and cruelty, depending on My whims; being pampered, pleasured, and amused; training males to become the perfect playthings they were intended to be. All men are animals, some just make better pets than others.

NAME: Mistress Appreciats, **CITY/STATE:** Around, **STATS:** A sweet beautiful blk domme female, **DESCRIPTION:** I love taking care of MY pets! Keeping MY pets on their knees, with a little help from (MS WHIP) remember, THE AZZ WILL SUFFER FOR YOUR DIS-OBEDIENCE. My foot in the middle of YOUR back. Bending you to MY WILL....Promise you "I will break you, or make you wish you were" "Gaze into my eyes, for you are about to journey to a place that only I know the way, TOTAL SUBMISSION, if this be your dream... Then hand ME your leash and follow ME"

A CHATROOM CHAT:

LovnSub:	Barbara. licking other foot now
BarbaraDom:	~~~~ LOL ~~~~
DomFem:	<—grabbing hand full of SubM's hair and pulling back
SUBM:	who do you want to IC you Bell
BarbaraDom:	Lovn working hard Laura hunny
Mistress Bell:	you
SUBM:	yes on my knees as you grab my hair
SUBM:	i feel your power over me
LovnSub:	Barb > and i am eyeing those high heels that you just took off
DomFem:	<—throwing you forward, fall on your hands and knees
BarbaraDom:	SubM?? hmmmm ~ u didn't wait for me??
LauraBeGood:	not hard enough youshould come over here
SUBM:	yes DomFem at your feet ass in the air
BarbaraDom:	Laura ~ is Lovn working you??

SUBM:	sorry Barbara
DomFem:	lift your ass up higher slave
LauraBeGood:	no only you
SUBM:	yes DomFem up all the way for you
BarbaraDom:	SubM—> grrrrrr
DomFem:	<— spanking SubM's ass with riding crop
LovnSub:	Barb > I would need two tongues
SUBM:	mmmmmmmmm...yes beat it
LauraBeGood:	that guy emailed me and couldn't even spell at all
BarbaraDom:	you wish u had 4 tongues don't ya??
SUBM:	sorry MS Barbara i owe ya
BarbaraDom:	who Laura??
LovnSub:	Barb > :-)~~~~
LauraBeGood:	i forward it to you
DomFem:	<smak
LauraBeGood:	To late SubM had your chance
BarbaraDom:	k
SUBM:	thank you misstress may i have another please
BarbaraDom:	oooh ~ Now what SubM... ya got Laura pissed
DomFem:	<smack
LovnSub:	Barbara > question?
BarbaraDom:	what?
DomFem:	<smack smack smack>
LauraBeGood:	miss DomFem?
DomFem:	<smacking SubM's ass harder and harder
Subby Pup:	ill do whatever anyone wants im me//////
DomFem:	yes LauraBeGood?
LauraBeGood:	give him 5 from me and 5 from Barb please
LovnSub:	Barb > saw a chat room when asked which women liked their asses licked, MANY liked it
SUBM:	mmmmmmmm,,,,,,yes nice & hot
DomFem:	ok
BarbaraDom:	lolololol ~ yah...he's a real smacked azzzzzz
DomFem:	smak
DomFem:	smak
DomFem:	smak
DomFem:	smak
DomFem:	smak
DomFem:	take that you filthy slave
LovnSub:	Barb > never saw that before
SUBM:	makeing me hotter

BarbaraDom:	so??? what ur question Lovn Sub???
Subby Pup:	i like to lick their asses
DomFem:	do you like it?
SUBM:	yes DomFem
LauraBeGood:	he didn't even thank you
DomFem:	answer me correctly slave
LauraBeGood:	SubM your worthless
SUBM:	thank you Fem
DomFem:	address me as Mistress Fem you worthless slave
BarbaraDom:	Can I Pee on you??
DomFem:	let me hear you
SUBM:	yes you can Barb
DomFem:	Barb would you like to pee on SubM?
LovnSub:	Barb > hmmm...I have seen that others like it
BarbaraDom:	lol :)
LauraBeGood:	she wasn't asking you
SUBM:	yes Mistress Fem
LauraBeGood:	Barb i'm watching
DomFem:	Barb, would you like to pee on SubM?
SUBM:	i'll take that too Barb :-)
BarbaraDom:	SubM... you don't know how to behave with a lady like Laura
LovnSub:	Barb > would you have said yes to that question some guy asked about ass licking
Recruiter:	ladies. interested in earning Good money from your home? IM me for info
SUBM:	i would in person for sure

There you have the world of the dominant woman. I find this lifestyle just a little on the silly side for me, even though the participants take it very seriously. Trying to figure out why a man would allow himself to be tied up by a woman who is wearing extra tight leather, allowing his testicles to be bound by rubber bands, and then hung from a pulley, is beyond me.

I have seen the pictures of the "horse" men—men dressed in the most bizarre outfits, but mainly naked, with reins attached to their heads, pulling a bare-chested, whip brandishing woman around the countryside in a pony cart. We won't even discuss what was attached to their genitals! All of this left me smiling to myself, shaking my head, and my standard thought rumbling through my head, "Do I know any of these people?" Actually, I will admit that now when I am driving through the country I am hoping to catch a glimpse of more than cows and horses on the hillside.

The dominatrix is adept at getting the right response from the submissive when she wants it, and we must not forget there are two types of submissives who seek the dominant female; the real time submissive, and the paying customer submissive. The real time submissive lives to serve his Mistress no matter what, and the paying customer goes to a professional dominatrix, pays a couple hundred dollars, goes through the torture and humiliation, and goes home. These are two different types of domination for the dominant female, and I'm not sure which one is preferable. In real time the dominant female always has someone to do the dishes and laundry, but in the professional dominants life, at $200/hr, she always has money to go shopping. This could be a hard decision for many women.

Donna Tracy

A LADY BY ANY OTHER NAME

The lady is a lady, is a lady, is a lady. Or, a "Communications Expert", or a "Cover Girl", or a "Personal Escort". No matter what you call it, there are a lot of business-for-hire transactions taking place in public Internet chatrooms.

In my quest to visit and capture the essence of the Internet chatroom, I have discovered the working woman no longer needs to reside on a darkened street corner in the cold night air. All she needs is a computer, a credit card machine, and some beautiful photographs of herself, or someone she can pass off as herself, and she is instantly in business from the comfort of her home.

Even though the men who frequent these rooms have no idea what these women really look like, or if they are even women, it was amazing how they responded to the women's come-ons. They would beg to have photos e-mailed to them from the women to download. Once the photos were downloaded these men would usually go berserk over the woman's looks, and whatever other attributes she was offering in her photos. I have no idea how large these men's hard drives are, but they should be full in three months time. Oh, and one other item I must mention—when I visited these women's web sites and saw their photos, well all I can tell you is, "One is more beautiful than the other!" "Wow!" Anyone of these women could be right from the Miss Universe pageant or the pages of <u>Playboy</u>, and to think they are sitting home on a Saturday night just to speak with all these Internet men. Now, this really does my heart good to think that there are women out there who would be humble enough, with their looks and bodies, to speak to the "average" man. (Average here means, you do not come with a yacht, a credit card at Tiffany's, or a private jet.)

The photos these women pass around like candy can be extremely sexually explicit, and are easily accessible to children. The young girls, ages 13-16 who happened to wander into these rooms would be asked immediately by some of the men to IC them and e-mail their photos.

What I did find extremely interesting were the amount of young women, 18-20 years of age, who are engaging in this type of solicitation. Most stated in their profiles that they were dancers, strippers, or adult models who were working while going to school. Some would only do phone sex, others would only have cyber sex, but there were a countless number who would travel the entire United States to meet with men—if the price was right. It is definitely "woman in control" in these rooms.

There is also another side to some of the women in these chatrooms—they just plain enjoy sex and there is no charge for any photos, cyber sex, or phone sex. All they want is a good time right now, and to quote several, "SEX, SEX, SEX!" which can make it rather hard on the working woman—what with all the freebies around, she needs to be on her back, phone, or computer, as the case may be, in order to pay the rent for the month.

Being rather of an "older" generation, I never realized how accessible sex was through the click of a mouse. It is truly a different world than thirty years ago. Who would have ever thought we could send a man to the moon and make computer sexual activity available in every home in the world in my lifetime? I believe the Internet chatroom has made a significant change in the sex industry today—after all, "Why pay for the cow when you can get the milk for free?"

Here now are some rather interesting and provocative profiles and chats taken from ladies chatrooms. I have also included profiles of ladies that I consider to be on the bizarre side, but that you can decide for yourself.

NAME: SexyBabe, **CITY/STATE:** Michigan, **STATS:** BiFemale, 5.1, 125, tan, tattoos, blonde hair, brown eyes, 34 29 30, **DESCRIPTION:** If u have no profile don't ic me as simple as that. Freaky sex I am looking to try k9 horse and what ever else comes along. Very open minded to all kinds of sex ic me if u want to talk. Very into golden showers and I love bi man mmmmm.

NAME: Let's Play, **CITY/STATE:** NJ, **STATS:** female, **DESCRIPTION:** I DON'T' MEET…AS IN NO MEETING! Sex is Evil Evil Is Sin Sin Is Forgiven So Lets Begin!! Pretty much into anything that gets the blood flowing. I am a Stripper & I am a super Star? Now guess what kind of Super Star? After you see who I am all you will ever think of is me and desire me more and more. I have pics of myself do you care to see?

NAME: CallMe4Fun, **CITY/STATE:** Everywhere, **STATS:** Female, Single, Bisexual, **DESCRIPTION:** E-mail me your 3 digit area code for a list of people in your area that want to meet you for a "GOOD TIME" & "PHONE PLAY" I will send you a list of people in your area code that WANT TO MEET YOU! Pictures are available. THESE ARE REAL TIME MEETINGS. Email Me Please us the link in this profile to email me at www.XXXX.com

I really want to know how the below lady has time to be a nurse with all the traveling she does. Maybe she is a "traveling" nurse?

NAME: PlayNurse, **CITY/STATE,** CO, but I travel to AL AZ AR CA CT DE FL GA ID IL IN IA KS KY LA ME MD MA MI MN MS MO MT NB NV NH NJ NM NY NC ND OK OH OR RI PA SC SD TN TX UT VT VA WA DC WV WI and WY, **STATS:** Female, single, nurse, **DESCRIPTION:** Play online, parties, rave, doin bad girl things. I have a adult website and im also a nurse. www.XXXX.com

NAME: Asian Beauty, **CITY/STATE:** Funland USA, **STATS:** Female, Single, and Bi Livin Lavida Loca, **DESCRIPTION:** I like Voyeurism, Deep Sensual full body hot oil massages, Howard Stern, naked twister, webmistress, chatting with new friends, bababooey, live web cams, kai, movies, dining out, comedy clubs. I am a dancer, voyeurism, adult asian film star. Love is sex, sex is sin, sin if forgiven so let's begin! Sex is lik math ADD the bed, Subtract the clothes, DIVIDE the legs, & hope that you don't MULTIPLY.

NAME: Kisses&Hugs, **CITY/STATE:** South, **STATS:** Wild, Kinky & Free Female, **DESCRIPTION:** Love Naughty fun, Love to explore all types of fantasies. Kinky, naughty, little girl fantasies, no boundries ~smile~ Adult Conversationalist ***21 and over only. Share your wildest darkest fantasy with me*** NO CYBER, FREEBIES OR REAL TIME MEETINGS so please don't ask

NAME: Lady With Cuffs, **CITY/STATE:** South Florida, **STATS:** Female, Single, **DESCRIPTION:** e-mail me at www.XXXX.com to find out what I like! You mean phone?...Yes, I do phone...Ask me... I am a Police officer. And, yes, in Florida, you can be a cop at age 18, just read the statues before you ask me the same question!

The following chatroom only contained links that one could click on for instant access to adult sex sites, which continuously scrolled in the chatroom. Of course I had to check them out and all I was asked was if I was 18 years of age or older—and click—I was in tits and ass heaven! I'm not sure if this one lonely, young soul ever got some cybering going, but I'm sure he understood what type of chatroom he had wandered into just as soon as a credit card was asked for.

A CHATROOM CHAT:

CLICK ME:	Sexy Bikini & Thong Pics...CLICK HERE
PoundKitten:	SexKittens Playroom ^...^
CLICK ME:	4 the Hottest Babes on the Net...Click here
CLICK ME:	Sexy Bikini & Thong Pics...CLICK HERE
CallMeAnything:	Check Profile & Email if you Wanna See

CLICK ME:	4 the Hottest Babes on the Net...Click here
CLICK ME:	Sexy Bikini & Thong Pics...CLICK HERE
PoundKitten:	SexKittens Playroom ^...^
CLICK ME:	4 the Hottest Babes on the Net...Click here
CallMeAnything:	CLICK HERE to see a Tall Hot Blonde~LIVE
CLICK ME:	Sexy Bikini & Thong Pics...CLICK HERE
CLICK ME:	4 the Hottest Babes on the Net...Click here
CLICK ME:	Sexy Bikini & Thong Pics...CLICK HERE
It'sChad:	any women want to cyber?
CallMeAnything:	Check Profile & Email if you Wanna See
CLICK ME:	4 the Hottest Babes on the Net...Click here
PoundKitten:	SexKittens Playroom ^...^
CLICK ME:	Sexy Bikini & Thong Pics...CLICK HERE
CLICK ME:	4 the Hottest Babes on the Net...Click here
CallMeAnything:	CLICK HERE to see a Tall Hot Blonde~LIVE
CLICK ME:	Sexy Bikini & Thong Pics...CLICK HERE

NAME: TooManyHormones, **CITY/STATE:** NY, Long Island, **STATS:** I just turned 19, 5'5-1/2", long shapely legs, busty (36D), blonde hair, brown eyes w/green around the edges, 114 lbs, **DESCRIPTION:** I have a video camera hooked up to my pc... What kind of pic do you want me to send you? Tell me how u want to do me...if your message makes me wet, I'll reply ... who knows...maybe we'll actually meet sometime and you can act out your fantasy. Tell me what you fantasize about. I like oral, anal & especially rough&forced watching pornos, partying, taking sexy pics, publically exposing myself ☺ Getting guys off... and of course having my cute lil behind spanked OTK!! I am a student... trying to become a model, or even adult model... I love watching porno's so why not star in them... mmm. I'd love it! "I don't care who you are or how old your are, just do me!" Us Nymphos can NEVER get enough!!!!!! Give it to me HARDER, and IC me and lets have fun!!!!!

NAME: Eastern Phone, **CITY/STATE:** East Coast, **STATS:** Female, married, 46 yrs old, **DESCRIPTION:** email me with age or www.XXXX.com. My likes are sensual phone chatting, fun on the beach, getting a deep, dark tan on my hot, sensual body. I am Self-Employed...Personal, Discrete Communications Business. I'm a hot, sensuous older woman. I love looking young for my 46 yrs. And enjoy the many looks I receive. ☺ I love to please a man in all ways. Why chat with a younger girl, when you can have an experienced, older woman!

This next profile I would venture to guess is for real. But, with my sense of humor I absolutely love her little ditty of a poem at the end. Of course it's not by Longfellow, but it could definitely be a bumper sticker!

NAME: Woman's Fury, **CITY/STATE:** Native of the ninth plane of Hell. Grew up in Castle Malebranche off the River Styxx. I now reside in the Gulf of Mexico, **STATS:** Female black widow, **DESCRIPTION:** I collect skulls, blades for decapitations, and venomous reptiles. I am a part-time hunter, full-time recruting officer of Hells Army. Otherwise…SELF EMPLOYED! HI FLOORMAT! "Time for lust…Time for lies…Time to kiss your ass goodbye!!!"

NAME: LikesEmYoung, **CITY/STATE:** New Jersey, **STATS:** Female, divorced & lovn it, **DESCRIPTION:** Playing and Teasing…Flirting…some Heavy Petting ;) and KISSING if you are a good BOY! Let Mommy help you Get Cleaned up…i can't have you at the dinner table that way! Phone does it for me…I am very REAL and love phone with you…i am Very Reasonable…but, NOT FOR FREE. Be a GOOD BOY sweetie and when you need to be BAD call MOMMY ;)

If anyone can figure out this next profile, please write and explain it to me. I think the Arizona sun has done a number on these two airheads. Men don't bother, but guys are ok. Just what the heck does that mean?

NAME: Becky & Rachel, **CITY/STATE:** Arizona, **STATS:** Females, we both live w/Becky's husband, **DESCRIPTION:** We love shopping, travel, spending money. We are bifems. Men: There's nothing you can do for us. Don't phone, don't cyber, so don't waste our time trying. We will talk to guys though.

NAME: SchoolDancer, **CITY/STATE:** Fort Lauderdale, South Florida, **STATS:** Female, single, long blonde hair and deep green eyes (36c-24-34) 114lbs 5'6" My sizes are 36c bra and size 5 panties, 21 yrs old, **DESCRIPTION:** I love studying the Male Anatomy & Experiementing, Sex, Sex, Sex, and More Ses, Wearing out men, Sexual interaction with very few limits ☺ ☺ Love to try everything. Sending my—nude-self pics. I love hearing fantasy's tell me yours and I will tell you mine. I'm an xotic dancer paying my way thru school…Everything's Better when it's wet!

A CHATROOM CHAT:

BillyBoy:	any of you bad girls feel embarassed about any lewd comments
LoveTalking:	like what Bill?
BillyBoy:	any sexual comments

LoveTalking:	heck no
HighRider:	"i would like to slide my tongue into your wet pussy and lick you til u scream"
HighRider:	is that lewd?
BillyBoy:	well some 18 year old girl reported me because of what i said to her
April Shower:	hmmmmmm
April Shower:	very lewd
LoveTalking:	slightly Rider
CoolBabe:	<<<<Likes the fact that some guys are hard in here
Sugar&Nice:	what did you say
HighRider:	ok, well i'll be good then and just say "hi"
HighRider:	Send me a pic and make me harder Babe
BillyBoy:	i was on a female cyber site and i said since she is a cheerleader she must have a great ass
PrettyFeet:	well this is a bad girls room LOL
LoveTalking:	that's far from lewd Bill
HighRider:	Well send me a pic and make me harder Babe
BillyBoy:	she needs to keep her ass off of those sites
Big14U:	cool looks like i'm the only fuck stick today lmao
LoveTalking:	ugh
Big14U:	awwww damn lol
DannyHung:	all hott and horny females wana chat with a hott guy with self pic press 6969 or ic me
April Shower:	fuck stick??/ hmmmm
Big14U:	need a snickers?
Backseat Lover:	32/m from massachusettes lookin for fun with a bad ass girl?
DannyHung:	all hott and horny females wana chat with a hott guy with self pic press 6969 or ic me
CoolBabe:	watching the men chat in here
CoolBabe:	*smile*
LadyTerrific:	sheeeeeeesh its quiet now
Big14U:	i killed it
LoveTalking:	:::belch:::
Big14U:	i think it was the fuck stick somment <shrugs>
LoveTalking:	i should paint my nails
CoolBabe:	No foreign objects in me
CoolBabe:	no no no
CoolBabe:	Not that kinky

Big14U:	paint mine too Love* <yawn>
Mikey69:	I wish some bad girl would ic me!
LoveTalking:	no no... you're supposed to paint mine
Big14U:	noooooo
LadyTerrific:	yesssssss
Big14U:	the toes maybe...
HighRider:	i'll paint your nails
HighRider:	and face
HighRider:	and ass

NAME: JustCall, **CITY/STATE:** Everywhere, **STATS:** Female, single and not looking, **DESCRIPTION:** You can call me anything, but Call Me!! NO CYBER, NO PIC TRADING, NO MEETING! Tell me your fantasies...I love to role play & enjoy all kinds of family fun. Hot phone talk—I have no taboos...What do you really want? A mommy? A daughter? A master? A slave? I can be all of these. Professional Phone conversationalist—Fantasies Fulfilled, Bottoms Spanked, Feet Tickled...check out the link below for info. Give me your ear, and I'llgive you everything you've ever desired. www.XXXX.com

NAME: WorkingWife, **CITY/STATE:** FL, **STATS:** Female, happily married, **DESCRIPTION:** Letting others see the Real Me. I'llshow you more than my Heart & Soul. Tell the truth, haven't you ever wondered what your neighbors wife really looks like? I could be the women living next door to you, and you'd never know it.

NAME: StudentOnCampus, **CITY/STATE:** WI, **STATS:** Female, 18 yrs old, college student, **DESCRIPTION:** I'm your little angel...<wink>. I have pics to show if you ask me for them! if you are going to email me, please put "info" in the subject or I will delete. College student & part time play girl...wanna hear my stories, then you have to call me ;) I do this to help pay my way through college. If you want to play for free, then you will have to find someone else....if you don't mind have good fun for your dollar, then I'm the one for you ☺~~~~~ NO CYBERING, MEETING OR TRADING!

You have just got to love these college students. They have come up with some of the most ingenious ways to supplement their college tuition for their families. Why they would ever need college with this type of entrepreneur quality, I have no idea.

NAME: SixAtCollege, **CITY/STATE:** We Travel AnyWhere ak al ar az ca co ct de dc fl ga hi id il in ia ks ky la ma md me mi mn mo ms mt nc nd ne nh nj nm nv ny oh ok orpa pr sc sd ri tn tx ut va vt wa wi wv wy, Boston, Detroit, Miami, NYC, Pittsburgh, Atlanta, Tampa, U name It, **STATS:** 6 Escorts, Single, Females, **DESCRIPTION:** We love to suck on Big Popsicles, Talk on the phone, chatting, touching, feeling and LICKING DA PEACH (we are bi not les) Our PEACH's Have NO FUZZ, love tattoos and body piercings on both men and women Email Us 4 Link! Escorts, work for Playboy parttime, strippers, lap dancers, Varsity cheerleaders, Babysitters, Sex Therapists. Our little black book is so thick it needs cliff notes…Sex is when a guys communication tells the girls imagination to increase her Plateau of Love Making, did u get the idea or do you need a demostration? Avi, jpeg, pics, we gott'em.

NAME: Lil Deb, **CITY/STATE:** Miami, South Beach, FL, **STATS:** 18, single, 5'8" (mostly my long legs), long blonde hair, 105 lbs, firm bust (34DD), **DESCRIPTION:** I have lots of naughty pics, I am VERY oral…I love hair pulling, spanking, really dirty, nasty talk, anal licking, face riding and tasting hard, wet cock. Enjoy talking to guys and trying to elevate/stimulate the conversation to new sexual heights. I love giving ~blow~ jobs…I love to be spanked! Wooden ruler, brush, prison strap, small round wooden paddle, yardstick.

I have become extremely jealous of the women who seem to be able to travel at the drop of a hat, from these chatrooms. It takes an act of Congress for me to be able to take time off and travel, but I guess that's the real difference between a "working girl" and a "working girl."

NAME: LoveMyCam, **CITY/STATE:** nj ny ca az co ct de fl hi il ma md nv nc pa tx, **STATS:** Female, 27, single & not looking!, **DESCRIPTION:** Some of you know me as XXXXXX ☺ You can see a few pics at that profile name but you can reach me on this profile ☺ TO GET INFO FROM ME Put "INFO" in the subject along with your AGE…ALL other subject mails WILL be deleted! Teasing, teasing and more teasing. Let's play on the phone or on cam! This hot little Italian loves SHOWING OFF for you! DON'T ASK ME TO CYBER! I'm here for your pleasure but I DON'T play for free. You can also IC me for my pics! I'm a Sexy Italian Phone Entertainer & Cam Star (wanna watch me?). I like being an exhibitionist, do you like being a voyeur? Pls DON'T ask me to meet; that may come in time. I don't like men who ask for things for FREE cause I'm NOT. If I wanted to cyber, I would but it's so BORING so if you want to play pickup the phone & call!

NAME: Used Shorts, **CITY/STATE:** Detroit, MI, **STATS:** Female, single, 5'8", 140#, 36D-28-36, Green Eyes and Brown Hair, **DESCRIPTION:** I am selling MY Used & Worn Panties! ~~~WARNING~~~I DO NOT CYBER SO

STOP ASKING!!!! PHSx YES! EMAIL TO INQUIRE ABOUT RATES! If it looks good, Sniff IT! ALSO ASK FOR A PIC AND A LINK!

The following chatroom chat shows how some men are on a quest to gather all the nude photos of the women who frequent these chatrooms. At first I thought "The Bestman" may have been "HaveUndies" pimp, but evidently from the conversation they have never met. Maybe it was just wishful thinking that she was already a major part of his life. Wonder if he would have anything to say if she turned out to be a he?

A CHATROOM CHAT:

HaveUndies:	sent the nude
The Bestman:	undies
TRUCKER:	undies did not recieve
HaveUndies:	yes Bestman
The Bestman:	i love you
HaveUndies:	ill resend
TRUCKER:	<~~~~tears in eyes
TRUCKER:	lol
The Bestman:	Haveundies = erection
The Bestman:	when is the episode gonna be on
HaveUndies:	thanks Bestman
HaveUndies:	mid sept
Susan4U:	HI room, I am HOT tonight
The Bestman:	i will be watching
HaveUndies:	send Trucker
The Bestman:	do you have more pics Undies naughtier ones
OutofState:	WASSUP HAVEUNDIES
Honk'n Male:	hey Undies, got the pics, thanks!
OutofState:	CAN I SEE A NAUGHTY PIC TOO
The Bestman:	no outofstate ask more politely
OutofState:	LIKE
The Bestman:	can i please see your lovely pics
OutofState:	LOL
The Bestman:	do it now
TRUCKER:	god Undies will you marry me lol
The Bestman:	I LIKE UNDIES
OutofState:	CAN I PLEASE SEE UR LOVELY PICS UNDIES?
The Bestman:	UNDIES WILL YOU SHOW THE GENTLEMAN

118

HaveUndies:	sure to see the pics press 7
The Bestman:	I WANT MORE
OutofState:	7
JohnnyBoy:	7
The Bestman:	HAVEUNDIES MORE
TRUCKER:	7
The Bestman:	MORE7777`
OutofState:	*7777777777777777777777777777777777777777*
Susan4U:	may look for another room if this one does not liven up
TRUCKER:	whats up with you SUSAN
The Bestman:	DONT GET TO HYPER BUDDY OR YOU WONT RECEIVE
TRUCKER:	lol let me know where you go
The Bestman:	UNDIES ARE YOU TAKEN
TRUCKER:	ic me when you do
Susan4U:	Looking for a hot room
TRUCKER:	its here in my house lol
Susan4U:	lol
TRUCKER:	your welcome to come over lol
Susan4U:	HA! and "do" what?
KingWonder:	art there any hot wet females
KingWonder:	in the room
Susan4U:	<—hot and wet
KingWonder:	well how wet
TRUCKER:	oh baby i'll pick you up
The Bestman:	I ALLREADY SAW THOSE BABY I WANT TO SEE NAUGHTIER ONE
TRUCKER:	even if i have to fly lol
HaveUndies:	that one you did not see dif pose
Susan4U:	Jest right and you?
The Bestman:	NOW I WISH I HAD A SCANNER TO SHOW YOU MY PICS

NAME: I'mAllAttitude, **CITY/STATE:** Everywhere, **STATS:** Female, Single, Trim, Attractive, **DESCRIPTION:** ...and baby, Talk dirty to me! I Want Actin Tonight, Satisfaction alright! No Profile-No Chat. I collect images of visual erotica. I collect lovers of both sexes. I seek real time meetings with people who stimulate and move me. Can you get my attention? Love exchanging Dogfart images (prefer earlier series)! Writer of Erotica. Open to new things. Have self images to exchange with right person. MEN: Get my attention or get lost. A hard man is good to find but a soft woman can beat him every time…Are you man enough for a lil thing like me?

NAME: LoveWatching, **CITY/STATE:** New Jersey, New York City, Staten Island NJ, NY, SI, **STATS:** Female, Single, **DESCRIPTION:** I love to watch people have S**. I would love to see you doing it. Please feel free to send me a video cassete of you in action. Either solo or with a partner. I like to watch while my fingers are fast at work. I am looking to collect all the movies I can. Please mail me any p*rno films you are no longer using. The sicker the better. Golden showers, beasty, SCAT, Gay, Freaks, Straight sex, Gang Bangs. I am a student and can not afford to maintain my devient habits so please help. Please check out my web site, there is a pic of me and my mailing address.

NAME: HornyHousewife, **CITY/STATE:** Ft Lauderdale, FL, **STATS:** Married Bi Female, **DESCRIPTION:** I am part of the SexyBiWives Club, we are a group of bi-wife "swingers" from South Florida that enjoy having our husbands take pictures of us posing and playing … you must check out more of us on our home page, click the link below and then check out our favorite links. If it turns you on looking at it, you know it turned us on taking it… www.XXXX.com

NAME: Let's Play, **CITY/STATE:** al ak az ar ca co ct de dc fl ga hi id il in ia ks ky la me md ma mi mn ms mo mt ne nv nh nj nm ny nc nd oh ok or pa ri sc sd tn tx us vt va wa wv wi wy, **STATS:** Female, Married ;), **DESCRIPTION:** Sex of all sorts—and I mean all—to the extreme—sometimes tender—sometimes hard—I love it all. Looking meet people and friends and men to do!! *grin. I like to play: Fetish, Housewifes, BDSM, S&M, Leather, Panties, Foot Fetish, Slaves, Domination, Lingere, sex, xxx—I am horny. Nothing good in life is for free—not even me—I cost ya exhaustion ☺ pics from me on my couch—videos too.

Somehow the following chat just doesn't ring true to my ears. If I had a nickel to bet, I would bet right on the nose to win that Cute Lori is a guy that is playing around in drag. Most of the women who frequent these chatrooms are more than eager to exchange their photo. Photo = money. Money = phone sex. Phone sex = everyone's happy!

A CHATROOM CHAT:

Mr. Richard:	got any pics lori?
Cute Lori:	no pic
Mr. Richard:	bummer

Cute Lori:	ya bummer
Mr. Richard:	how old are ya?
Cute Lori:	26
Hot4U:	any ladies in here got some pics
HoneyBuns:	hi 19/f, ic me
HankerN4IT:	HI LADIES IC ME FOR CYBER SEX ALL NIGHT
Cute Lori:	a/s/l underwear check
Heartbreaker:	23
Mr. Richard:	29/m/FL none
Cute Lori:	26/f/cali pink thong
Mr. Richard:	Does any women in here have a webcam?
Heartbreaker:	are you hot lori
Steve1976:	Lori
Cute Lori:	ya im hot
Steve1976:	Lori- why no ic's??
Cute Lori:	i get to many
Mr. Richard:	She's 400 lbs of sweatshop hot
Mr. Richard:	lol
Mr. Richard:	just joking
Cute Lori:	lol
Cute Lori:	i bet your hung
Mr. Richard:	I am
Cute Lori:	lol
Mr. Richard:	you think that your cute huh?
Cute Lori:	all man on line r hung
Mr. Richard:	but of course
Cute Lori:	and all women r fine
Mr. Richard:	sure
SLCKITTY:	hello room
SLCKITTY:	whats happening tonight?
Mr. Richard:	k, bye all.
Cute Lori:	but i really am fine
Mr. Richard:	sure ya are
Cute Lori:	ok...bye
Mr. Richard:	lol...
HankerN4IT:	HOW MANY PUSSIES R WET IN HERE
Cute Lori:	and your really hung
Cute Lori:	hehe
Karen8221:	17/f
Mover:	hung on ur pussy
Cute Lori:	i dont think mine is wet yet

121

Mover:	hi 17 f
Cute Lori:	let me check
Mover:	Karen
HoneyBuns:	19/f here, ic me to chat
SLCKITTY:	thats what i want to know is how many pussys are wet tonight?
HankerN4IT:	REALLY SO MAKE IT WET LORI
Cute Lori:	no not yet
Mover:	they need guidance
Cute Lori:	u make it wet
Mover:	to get wet
Mover:	hell yea i make it
Mover:	wet
HankerN4IT:	OK IC ME
Mover:	i lick it wet
SLCKITTY:	any BI women here?
Cute Lori:	i like a guy that knows how to use his tongue
Mover:	thats y god put them
Mover:	on men lorie
HankerN4IT:	I DO
SLCKITTY:	26 female bi salt lake city
Cute Lori:	thats what there for huh
Chipper:	Lori u got a pic?
HankerN4IT:	YUP
Mover:	u kno it
Cute Lori:	um...
Cute Lori:	no pic
Mover:	can i get to the licking
Mover:	sorry no scanner
Cute Lori:	lick your srceen
Mover:	my screen is now wet
Mover:	ya right
Mover:	she all pussy
Mover:	she a girl
Mover:	get no pussy say that
Mover:	teen girl ulike it anal
Cute Lori:	whos a guy...whos a girl in here
Network:	28/m
HoneyBuns:	hi, anybody wanna chat? ic me
HankerN4IT:	M
Mover:	im a guy
Philip5984:	42/m

| Vvega44: | guy/39 |
| Cute Lori: | im a girl |

NAME: Role Player, **CITY/STATE:** ConstantState of Arousal, the Impending City of Desire, **STATS:** Female, single, **DESCRIPTION:** Having great conversations with inventive Men and Women. Exploring the realms of fantasy and role play, creating a place for you and me in my mind…Adult Goddess, Smooth Operator, Home Communications Specialist—that's it! ☺. NO FREE phone, NO IDLE chit chat NO I DO NOT trade pics. The best thing in life are NOT always FREE.

NAME: YourFantasy, **CITY/STATE:** Buffalo, NY, **STATS:** Female, **DESCRIPTION:** I DON'T' MEET, I DON'T CYBER EVER! FANTASY'S, ROLEPLAY, DEVIANT ACTS, I AM IN UR IMAGINATION, I AM UR FANTASY UR DESIRE. I AM NOT UR GIRLFRIEND, NOT UR WIFE I AM EVERYTHING THEY CAN'T BE BUT U WANT. IC ME. I ALSO LOVE TO BE THE DOMINANT!!! 21 AND OVER ONLY! NO FREEBIES SO DON'T ASK. THE BEST THING IN LIFE IS ME AND I AM NOT FREE ☺ www.XXXX.com

NAME: Talk2Me, **CITY/STATE:** City of Passion, State of Ecstasy, Country of Total Pleasure, **STATS:** Female, Single, 24 yr old, 39 24 36, **DESCRIPTION:** Love Leather Have All Kinds Of Leather clothing, Boots, Love my men to wear leather and admire me in it. I wear it well. I collect men who wear it. We play with leather toys. Wanna play? Adult Phone Therapy. No it is not free don't ask NO FREEBEES, NO CYBER, NO OFF LINE MEETING. Hot tubs, music, wine, lots of playing, Yes, I discipline very well. Call me and let's be romantic as I fulfill all your fantasies the ones you are too shy too talk to anyone about…i am waiting.

NAME: PleasurePhone, **CITY/STATE:** Anywhere you are, Lover, **STATS:** Female, **DESCRIPTION:** Enjoy bringing pleasure and pain by phone…and believe me when I say I aim to please. Whatever it is you want, I've got it. No fantasy is too risque. Telecommunications specialist <evil grin> Sorry lover, no cyber, no freebies, no offline meetings, no idle chat. Why walk around in agony! Call me up and name your pleasure…I'll blow your mind and whatever else you like. www.XXXX.com

NAME: YoungOne, **CITY/STATE:** Phone Land, **STATS:** Female, 18, too young to be married, **DESCRIPTION:** I'm being sweet, naughty, submissive for YOU—Adult Phone Conversations, Exploring My Newfound Freedom, men, women, beach, cheerleading, writing stories…lots more—use your imagination…I use mine! Ssssh…I don't Want Mon & Dad to find out. Can you keep a secret? It has to do with my voice, My Naught Mind, And The Phone. "Good Little Girls Go To Heaven… Bad Little Girls Go EVERYWHERE!!!!" www.XXXX.com

NAME: Family's DeeDee, **CITY/STATE:** Never Land, **STATS:** Female, Single, **DESCRIPTION:** Having sex with my brother and other family members www.XXXX.com, I'm a nudist also. Stay close to the family (family sex)

The below chatroom chat features DeeDee from the above profile. It's nice to see that people will draw the line at some things in life. DeeDee found out she had better go and peddle her wares in another chatroom as these occupants just weren't buying what she had for sale!

A CHATROOM CHAT:

Family's DeeDee:	My bro. & I have finally put our INCEST Pics. Online
Loverboy:	oh STFU DeeDee
BADASSET:	I LOVE TO BE REALLY BAD
DREAM GIRL:	DeeDee, get that out of here
DREAM GIRL:	NOW
Loverboy:	oooooooo…haven't heard that one for awhile
Bruce Here:	any females for phone sex ic me
KATHY JOY:	DeeDee…they do have famiy fun rooms…GO THERE
Family's DeeDee:	oh loverboy you know you like it
BADASSET:	PHONE SEX IS THE BEST
KATHY JOY:	Bitch, get off of him…
Button:	too bad there is no video in the room
Loverboy:	yeah…<yawn>…right DD
Loverboy:	oooooooobaby
LadyWed:	:::::shudder::::::: why would folks get into family fun anyway?
LadyWed:	just don't do.
Loverboy:	go hump a cooter DeeDee
KATHY JOY:	omg…lol Lover
Loverboy:	lol
KATHY JOY:	she prolly would
Loverboy:	I'm sure she would
BADASSET:	HOW MANY GUYS IN HERE WOULD WEAR HIS GIRLFR PANTIES

PoorMe:	hello 23 m, lover of muff diving
BADASSET:	UM LOVE TO HAVE A MUFF DIVER
PoorMe:	here I am
BADASSET:	DO YOU DO NO FUZZ ON A PEACH ONE??
LadyWed:	Wench's Rules for the Road...the more a guy brags about how good he is at oral favors, the
LadyWed:	worse he is in reality.

NAME: LoveToTry, **CITY/STATE:** California, **STATS:** Female, Single, 5'8", 105 lbs, green eyes, blond hair, bra is 36C and I'm 22 yrs old, **DESCRIPTION:** Enjoy flirting, having fun with guys, if you kiss me don't get me wet, use your tongue and make it nasty, any kind of kinky nasty sex, perverted sex with random guys, wild sex orgies with my girlfriends, I love to give and receive head for hours! Yes I have a picutre. I work in an office doing temp work! I can bend in strange ways, what's your favorite position?

NAME: WantsAPuppy, **CITY/STATE:** Atlanta, GA, **STATS:** Female, Married, 26 yrs old, **DESCRIPTION:** Looking for an experienced BBW K9 woman in my area, age 25 to 35, STRAIGHT/SINGLE for friendship only < NOT sexual...I'm not bi > who has a large dog and interested in K9 for fun. Must be discreet, my hubby not joining. Interested in friendship aside from the K9 part. Have many interests. NO MEN... IF YOU IC ME <YOU WILL BE IGNORED! Prefer meeting someone who already has large K9's please.

I guess by now you have come to the same conclusion as I have—that things have certainly changed for the working girl since the computer age came on the scene. What with the invention of faster modems, faster processors, web cams, and lower interest rate credit cards, this profession has taken on a whole new meaning. It is amazing how many women (or men disguised as women) are participating in this occupation on the Internet. It was only when I actually became serious about doing research for this book that that realization hit me. I now understand why people would choose to be involved in the sex industry. Sex equals big business. Big business equals big money. Big money equals end of discussion.

THE COUPLES

A more fitting title for The Couple or Swinger lifestyle, would be "Exhibitionist", from my point of view. These people love to perform not only for each other, but also for or with others in the same room. While the majority of couples prefer to welcome a third partner into their world, that third partner is usually a bi female who only has sex with the female. It seems the husbands, or boyfriends, remain "true" to their partner by only watching the cavorting of two women. Extra men partners are rarely accepted, unless it is a total partner swap, as straight is the way for the majority of these men, and somehow they prefer not to see their wives/girlfriends with another man unless they are with another female.

There are countless swinger clubs in every state for the swinger lifestyle enthusiast. These clubs have applied very strict rules for membership which usually include applying for membership, a screening process, an interview process, and finally, if accepted, an invitation to come to a get-together to meet and get to know other members of the club. Most of these clubs never admit single men, but single women are encouraged to come out to "play" on any of the nights they are open. These clubs serve drinks, food, have rooms for sex, and may be located in a nice residential neighborhood. One club that I found was having a New Year's Eve party at a local hotel, and everyone was invited. Of course clothing was optional.

Now, that last sentence makes the statement of why I am not a swinger. If anyone thinks I am going to take my clothes off in a fully lighted room with a bunch of people around, well they had better think again! Some bodies should be kept under raps, preferably fur raps <gbg>, and I admit mine is one.

I found the occupants of the couple's chatrooms to be very nice, everyday, fun people. They kidded, joked, and seemed to have a good time just knowing each other and discussing the different clubs they have been to. Their photos on their websites were of people just like you and me, normal (well I like to think I'm normal, whatever that may be). Some were young couples, middle age couples, and older couples. They discussed their families, kids, and vacations with other couples in the chatroom, and they were very friendly to single women who came in the chatroom.

I did find it quite fascinating that the majority of these couples were comprised of a bi female with a straight male, and that the male partner had absolutely no problem with seeing his partner enjoy a woman. The rest were partners who swapped completely (two couples exchanging partners), with the less popular couple scenario being one that accepted a single male into it.

As a result of venturing into this lifestyle I can proclaim that I am not the only voyeur in the world, for this lifestyle from what I have seen is comprised mainly of voyeurs. I just prefer to do my observing on the Internet with my clothes on.

NAME: Frank and Pam, **CITY/STATE:** Anywhere, **STATS:** Married, **DESCRIPTION:** Interested in meeting other couples and select males (married or single) for ADULT fun. Age, race unimportant although black most welcome. We are very much into MFM but both can be bisexual with the right people. Love having others see me naked and perform for them. Love being with two men at the same time. I'm into mild bondage, humiliation, and enjoy letting my husband see me stick my tongue "where the sun don't shine" on another man.

I certainly hope the below partner finds his/her true calling in life. It must be terrible having so many professions at one time. ☺

NAME: Robert and Brenda, **CITY/STATE:** Raleigh, NC, **STATS:** Married, **DESCRIPTION:** Amateur gynecologist, vocal cord tuner, pipe layer, (public) hair stylist, french maid (specializing in those hard reach areas), and proprietors of "CREAM N' WEINIES". You have to get up pretty early in the morning to catch me peeping through your bedroom window.

NAME: Dez and Ida, **CITY/STATE:** Omaha, Nebraska, **STATS:** Happily Married and enjoying others, **DESCRIPTION:** Sex, Cples, Swing Parties, Bi-females. Select well hung men. She is Bi (loves the ladies) Hes STR8...SWINGERS!! It's a full time pleasure. Its our duty to take care of that booty. www.XXXX.com

NAME: Bobby and Karen, **CITY/STATE:** Ft Myers, FL; Carribean West Indies & Mid West US, **STATS:** Her 45, 5-7" 130# Bi, 37C/25/35, Lt Blk—He 55, 5-10", 160# Str8, Wht, **DESCRIPTION:** Sex, nascar, sex, Mtr Cyl's, Sex, House Parties, Hott SEX! Photos 4 cpls—Send 2 Receive. Both are professional by Day—Nite "Well Cum Find Out!" "Get It While It Is Wet - - Ride It Until It Is Dry!"

Someone needs to explain to me how one can be "straight" and still have sex with the same genre. I encountered a lot of this type of thinking in my travels, and I would still classify anyone who has sex with the same genre as either bi or gay. Am I wrong in my thinking here?

NAME: Brian and Honey, **CITY/STATE:** Plymouth Meeting, PA, **STATS:** Engaged Soulmates, **DESCRIPTION:** Sensual adventures / both straight / share and share alike / MFM, couples, she might experiment if the mood was right / fmf also a hobby. We are pleasure seekers...the rewards are awesome / prefers hispanic men...7"+, single guys... send a pic and an idea of our second meeting / 1st is a get to know each other.

I hate to tell the below couple, but what they are looking for in a man is what most women have only dreamed of, or read about in magazines. Good luck, you might want to try a fairy tale book—with pictures!

NAME: Bob & Tammy, **CITY/STATE:** Cleveland, OH, **STATS:** Single White couple, **DESCRIPTION:** Voyeur single couple wishes to watch a very sexy gorgeous petite fit white woman (first time encounter with hung black preferred) with a very muscular extremely endowed super hung stud (10++" X 7" around) especially BLACK. ONLY interested in watching a very sexy gorgeous petite HOT white female with a very muscular huge hung man. We ONLY WATCH the action...We do NOT join in. Looking for a very sexy woman interested in HUGE; and studs who are at least 10""X 7""around; Black men preferred but will consider others that measure up. YOU MUST SEND YOUR PIC TO BE CONSIDERED WITH PROOF!!!!

Here is a friendly chat taking place, which includes trying to entice a female to hop a plane and come play.
Any bets on whether they connected?

A CHATROOM CHAT:

Judy:	dang, i want to play...
Mitch & Iris:	come play with me Judy
Linda:	what do you want to play Judy??
Judy:	woohoo mitch,
Judy:	if only i could
Mitch & Iris:	this weekend maybe Judy
Judy:	lots of things Linda
Judy:	Mitch, i work weekends
Judy:	fri...sat...sun
Mitch & Iris:	fly in sunday nite
Judy:	and anydays i can pick up overtime
Mitch & Iris:	we will pick you up
Judy:	will have to check flts, as i go standby
Judy:	and i dont want to get stranded in phoenix

Robin:	what did i miss
Karen:	where are all the bi- gals at?
Mitch & Iris:	that would be good judy
Linda:	Here I am Karen.
Robin:	bi fem here
Judy:	Robin, you keep saying that
Robin:	i do?
Judy:	lol
Mitch & Iris:	everyone wants to play with Judy
Mitch & Iris:	including me
Judy:	mitch, well come on up here
Judy:	and play with me
Robin:	why haven't we done something about it
Mitch & Iris:	we are the 20th
Karen:	why is it that all the lovely ladies live so far from me?
Mitch & Iris:	we are coming up to meet sue
Linda:	where are you Karen
Robin:	where are you
Karen:	out here in the VAlley
Judy:	woohoo mitch, care to celebrate my birthday
Karen:	about an hour from you
Linda:	Oh Bummer
Mitch & Iris:	okay judy
Karen:	but ill be in VEgas tomorrow!
Linda:	I wont be
Robin:	Nice profile Linda
Judy:	mitch, you can be my birthday present...
Judy:	robin, i am in san fran
Linda:	thanks Robin
Mitch & Iris:	moreno valley here
Robin:	well we just might have to cum n visit
Mitch & Iris:	yeah im Judy,,,,,,present
Judy:	Mitch, do i get to unwrap you
Mitch & Iris:	slowly Judy
Mitch & Iris:	very slowly
Judy:	or will you just have a big bow on
Robin:	boy, i got on just at the right time, bi females are on
Robin:	yummy
Linda:	Bi Female Here Here
Mitch & Iris:	bi Perk

Karen: bi bi b i bi bi bi bi

NAME: Gary and Dolly, **CITY/STATE:** MD, **STATS:** Bi 5'4", 110 blonde, brown eyes, 34C-24-34 And, I don't meet alone! Single with Boyfriend. **DESCRIPTION:** BiF, Sex, Watching Sex, Being watched, Toys (will call to verify over phone) Looking for other BiF to join us or like cpl with BiF = No swap the men can watch! Looking for real time (NO fakes or Single Men) e-mail and send a pic to receive! So many women, such little time!!!

Many of the swinging couples have their own website. I found these sites very informative for someone so naïve as myself. It seems this is a fun lifestyle in comparison to the dominant/submissive lifestyle. The majority of these couples, except for swapping, are not into anything too excessive in the bedroom, living room, or kitchen, as the case may be.

NAME: Trucker and Jolene, **CITY/STATE:** Chicago burbs, **STATS:** Very attached, **DESCRIPTION:** Looking for fun sexy attractive 30-40 something professional couples. She likes her men tall, clean cut handsome and women very feminine and bi. If you think you fit write us, we are real and looking for sincere couple. Our bedroom...your bedroom. NO SOLO MEN. Got pics on our attached web page. Please send yours. Yes we like to know who we are chatting with. www.XXXX.com

NAME: Stewart and Cheryl, **CITY/STATE:** San Diego, CA, **STATS:** Latin cpl, married and sharing! 29/bi-female, 29/straight male, **DESCRIPTION:** Threesomes, and making love with others while my hubby watches. SAME ROOM SEX, 3SOMES & 4SOMES. Looking for young sexy couples, bi-females or selective sexy males. Single males we will contact you. Please do no bother US. She is: sexy 29 yrs young latina bi-female 5-4 120lbs long hair, sexy legs, nice body in great shape. MALES I don't play alone! He is: sexy 29 yrs young (shaved) bald head latin male, 5-11 205lbs sexy smile great shape muscular and musculine. NO PROFILE-NO CHAT SEND PIC TO RECEIVE, AND PLEASE SEND A REAL ONE. (LOCAL PEOPLE ONLY, ONLY) www.XXXX.com

The following profile is not a couple's profile, but actually an ad that I found on a person's profile for a private club in the hills of Topanga Canyon outside of Los Angeles. These types of clubs are located all over the United States, and I have included a link in the appendix that lists many cities with club locations, along with contact information.

NAME: THE CARNAL MANOR—Please visit our web site: www.TheCarnalManor.com, **CITY/STATE:** Topanga Canyon, California, **STATS:** A private and chic party house! **DESCRIPTION:** By invitation and reservation only, the party is for lifestyle couples and single women exclusively and is held on Saturday nights from 9pm until 3am. Please send E-mail or IC us for further information. Providing a refined night of frivolity for all of our party guests. Every sensual amenity is meticulously looked after and when you enter our doors expect to experience a grand night of erotic decadence, combined with a wickedly fun time! Web Page: www.thecarnalmanor.com

Have you noticed how many rules come with this lifestyle? These couples know what they want in other partners/couples, and they definitely set guidelines to be followed by the responding couple. Maybe we should all take heed when looking for a wife or husband?

NAME: Steve and Cory, **CITY/STATE:** Hinsdale, IL, **STATS:** Happily Married! **DESCRIPTION:** We are looking for people that live near us...if you are out of the area...No thanks! We are looking for couples in late 20's thru 40's!!! WE are NOT looking for singles so if you are, please leave us alone-we are in early 30's looking for Married couples. We are into full swap—but not into any pain or any exhibitionism, and No "curious (but not sure)" females!!! We do not teach! If you are a bifemale or a MARRIED (each other) couple with a BiF near our age IC or e-mail us...Wife is passive Bi...NO Singles guys!!! And NOBODy under 21!!!! We are fairly open-minded but have ZERO desire to be with singles...

NAME: Jay and Cindy, **CITY/STATE:** Arcadia, Pasadena, Temple City, CA USA, **STATS:** Male and Female, **DESCRIPTION:** S2R photo, Please G-rated first. No profile, no pic, NO Chat (Serious People), No phonies, Give Us respect and We will U (NO means NO). We are looking for discrete, Cples, Females ((Males send pic NO ICS)) Unless We ask you!!!! (Between 21-50) NOT into pain. We are Very Real!! Open minded Bi attractive (Not super models) cple Down to Earth (WYSIWYG) Seeking people within 30 mins Radius (sorry flake factor) Please NO PhoneSex or No cyber, We r lookin too meet new people who enjoy the same interests, We love to watch, Be watched, Full swap w/others in same room sex, As well as exploring new Adverntures.

A CHATROOM CHAT:

Missy:	<~~~dancing in my chair
JOE:	< has mink lined cuffs...
Anne:	<~~~~dancing

Anne:	Joe cum4me
JOE:	Missy- dirty or naked dancing :)
Ken & Dawn:	Hopefully both Joe
JOE:	<- not greedy...
DAVID:	go bone youre own girl ya jerk
JOE:	just loves the "view" :)
Ken & Dawn:	Carefull joe if you are regaining your sight you might want to take it easy. LOL
JOE:	hmmmm... seems like i am a stud, since some in here have NOTHING swinging, between
JOE:	their legs...lol
Anne:	lol...
Anne:	<~~~nothing swinging...
BOB & ANITA:	LOL JOE
Missy:	Anne send me the dear penis one again...I lost it :(
BOB & ANITA:	how can you loose a penis Missy???
Missy:	David, are you talking to me?
Ken & Dawn:	Maybe it was detachable...
Missy:	lol Bob...good question
JOE:	plastic, vinyl
Missy:	Got one of those Joe :)
JOE:	no kidding :)
Anne:	hey Joe when i get to NC think we might have kinky sex
JOE:	might?
JOE:	Anne- unless i'm DEAD
JOE:	the answer is YES!
Anne:	k ;)
JOE:	<- loves a good wife :)
BOB & ANITA:	cum to MD Anne letsfuc
JOE:	in "jeans"
Missy:	kinky sex
Anne:	i love it when you talk dirty to me Bob...
JOE:	anyone know the CURE for the braindead
Anne:	no Joe likes daisy dukes...
BOB & ANITA:	mmmmmmmm letsfuc
Missy:	I think you need a good f*ck David...might put you in a better mood
JOE:	Anne- you're reading my mind :)
Anne:	oh my... lol...
Anne:	just waiting for dickride

NAME: Tim and Leslie, **CITY/STATE:** Charlotte, NC, **STATS:** Married, **DESCRIPTION:** Do not send pic unless I ask…and G only. Must show face. Only play together. Under 45 only and *******SLIM********* No cyber No phone…real thing …NO BBW'S…Race unimportant looks are. Very openminded…wink wink. Professionals…3somes, bi, couples, hung, sex, male, female, play, swingers, lifestyle. No profile-no chat…no pic-no meat…be real or be gone…never send pic first so don't ask, but we have them, …We like to know who we are tradin with though…we all bang each other till we all come…variety really is the spice of life.

NAME: Ed and Linda, **CITY/STATE:** Chicago, IL, **STATS:** Married and very happy and secure, **DESCRIPTION:** Wife loves to dress very sexy. Watching, being watched, fondling, dirty dancing at CC&CP, not full swingers…Both very smooth where it counts. Both Professional. No singles or cheating husbands "Thank You Very Much"

Below is an engaged couple that is already looking for extra curricular activities, even though they are not yet married. I always thought such activities came after about 2 years of marriage. ☺ I am learning something new all the time!

NAME: Tom and Mickey, **CITY/STATE:** Newport Beach, CA, **STATS:** Very good looking engaged couple. Her Brn/Grn 5'7" 115lbs (Bi curious/some experience) Him Brn/Grn 5'10' 175, **DESCRIPTION:** Look for erotic fun with attractive "real" So Cal Cpl. Only interested in people who want to meet or talk to us both together. We don't meet without each other. We have several real pictures we can send if you have same. (Both very Attractive) No posers/phonies!We like to be watched and watch other attractive cples/bifems. NO SINGLE MEN—PLEASE DON'T ASK!

NAME: Howard and Karen, **CITY/STATE:** Northern Calif, **STATS:** Mid 30's, happily married, **DESCRIPTION:** We are into dining out, movies, dancing…camping…and good friends…W/A twist wink! We are not skin and bones so if that is what you are looking for,,,you will have to look elsewhere… ☺ We are both in very public professions…so being discret is a must! No Cyber…Bores us to tears.

NAME: Jerry and Diana, **CITY/STATE:** WA, **STATS:** Married, both STR8, **DESCRIPTION:** Meeting couples for fun or discreet sex. Sex is best…but two or more couples is much better…life is short so let play (sorry but no profile no chat) www.XXXX.com

NAME: Curt and Shelly, **CITY/STATE:** NY, NY, **STATS:** Very married, 28/F and 31/M, **DESCRIPTION:** We love traveling, living out all our fantasies, looking to have fun with people our own age who are attractive, fit and a little crazy. We often take weekend getaways. We just went to Hedo and cannot wait to get back!!! Please NO SINGLE MEN....Thank You! Professionals (or at least pretend to be). We travel often!!

NAME: Ryan and Peggy, **CITY/STATE:** Chicago, IL, **STATS:** Married, **DESCRIPTION:** Bi-females, couples, & even lesbians (if you want that duel pleasure)... No Single Males! We love sex... No anal!!!! No pain!!! No Single Males. "This has got to stop not the passion us not being able to resist each other... not you deep in me bodies thrashing drinking love from each other not being able to get enough of ourselves... but i pray it never stops'—Insatiable

Strip Pool! Now why didn't I think of that when I was twenties? This game could really catch on. If a gal has big boobs, she's a sure fire to win if she just keeps bending over the table for them long shots!

NAME: Hal and Martha, **CITY/STATE:** Tampa, FL, **STATS:** Married and still interesting!!! Him 5'9" 175# dark Italian / Her 5'4 size 8 blond hair green eyes, **DESCRIPTION:** enjoy naked hot tubbing...interesting COUPLES...enjoying life b4 we too old...ever play strip pool??? Professional, sexy, playful couple that likes to add a little magic to their life...we'd rather die while we're living, than live when we're dead.... www.XXXX.com

NAME: Paul and Cindy, **CITY/STATE:** SLC, UT, **STATS:** Married for 8 years, M 29 5'11" 195 pds very fit, F 28 5/3 115pds, **DESCRIPTION:** Wife is Bi, hubby is open, just like to have fun, be friends have great times no pressure—Tired of all Losers on Internet. WE r non smokers u be 2, discreet and clean (we r very real). Husband talks mostly, wife shy likes to be conservative and take her time, but watch out—Treat her right!!! Do not waste our time, we r not desperate. All u single guys, BYE

This couple's profile not only advertises for anyone who would like to play, but also for computer lessons. Guess we could call someone a "swinging teacher".

NAME: Willy and Grace, **CITY/STATE:** Louisville, KY, **STATS:** We are a couple who loves to play, **DESCRIPTION:** Love single men, women and couples. To get our pic you MUST send first. We play but only after Willy screens EVERYONE. This a firm rule, don't EVEN ask for us to break it. If you need to learn more about computers, I offer lessons in your own home on your own computer. Willy is bi male and Grace is Bi curious.

NAME: Vance and Michelle, **CITY/STATE:** Providence, RI, **STATS:** 32 Fem and 37Male, married white couple, **DESCRIPTION:** If you are a single male, please do not waste your time or ours. This also goes for married men who cheat on their wives or those with no profile. WE are looking for like minded couples close to us for friendship and??? No Cyber or Phone! Don't ask for a pic or what do you look like as your opening line. Everyone here is wondering what its like to be with someone else!

Yes! We finally hear from the "older generation", and yes, they are also swinging. Oh, this is terrific to know. And you kids thought you had the market cornered. Did any of you ever think about your own parents swinging? Kind of puts a whole new perspective on the idea, doesn't it? ☺

NAME: Ron and Jen, **CITY/STATE:** Las Vegas, Laughlin Area, **STATS:** A very committed, respectful, loving, sexy, senior couple—50s, **DESCRIPTION:** NO SNGL MEN! We enjoy ROMANCE, travel, RV, horses, boating, meeting friends, committed/REAL cpls 40 and over, FRIENDS FIRST. Lady of house is curious; gentleman is str8. Pls NO cheaters/wife or hubby—too much pain. Not to play on sight; but to establish a friendship first. Professional business owners. If you are traveling to Vegas, pls we are NOT tour guides. The couple that plays together, stays together. No cyber, PLS. We are mannerly— (respect and understand limits), non-pushy, and expect the same. And, PLS, no profile = NO chat. WE are interested in safe & considerate cpls for quality play N/S—D/D free.

A CHATROOM CHAT:

DJ & Sue:	hubby works graveyard so will be alone
DJ & Sue:	just me and my fingers
KAREN:	new to this looking for same room sex
KAREN:	bold but oh well
DJ & Sue:	did you get a profile Karen?
KAREN:	yes
KAREN:	24 hrs the com said and it would be done
Cowboy & Mrs:	WE WANT TO DO SOMETHING
Cowboy & Mrs:	TONIGHT
Cowboy & Mrs:	DAMMIT
SexySwinger:	Good Evening room, how is everyone!!
Nick & Nicole:	we see your profile now Karen...

135

Matt & Rose:	Hey Sexy Swinger
SexySwinger:	Hi Matt & Rose
Nick & Nicole:	2 on 2 works for us :)
DJ & Sue:	who is this WE??
Cowboy & Mrs:	Bored
Matt & Rose:	We still haven't found a couple that can "keep up with us" on the web cam!
DJ & Sue:	same here
Kevin & Liz:	lol Matt
Matt & Rose:	just haven't met the right couple on here yet, Kevin
Cowboy & Mrs:	SO-CALI COUPLE LOOKING TO GET INTO SOMETHING
Cowboy & Mrs:	ANY SUGGESTIONS PEOPLE
Kevin & Liz:	we have a hard time meeting anyone on the net have better luck at dances or clubs
Matt & Rose:	Not on a thursday night
Matt & Rose:	Yeah Cowboy
Nick & Nicole:	Same here clubs are the REAL way to go
Matt & Rose:	there you go
Matt & Rose:	you guys might get lucky with a stripper at a club
DJ & Sue:	is this Jerry or Dawn?
Nick & Nicole:	we been on the net for about 4 years and it wasn;t always this tough to meet folks
Matt & Rose:	If you're good tippers! lol
JERRY & DAWN:	both here
DJ & Sue:	ok
Karen:	hi any one who can give leads
SexySwinger:	<— stripper, lol
KAREN:	is there a pro in here
SexySwinger:	DJ, how are you tonight?
DJ & Sue:	I am ok hun
DJ & Sue:	what is your name Swinger??
Cowboy & Mrs:	so-cal couple here
SexySwinger:	Name is Jennifer, DJ
DJ & Sue:	you a couple??
Cowboy & Mrs:	wus up
SexySwinger:	Yep, hubby is in background, working
KAREN:	hello
DJ & Sue:	ok and his name??

Cowboy & Mrs:	were trying to get into something tonight but can't find anything to do
SexySwinger:	Duane, he's Australian
DJ & Sue:	cool
KAREN:	hey are you sall preoccupied
DJ & Sue:	nice to meet you Jennifer and Duane
SexySwinger:	Nice meeting you once again DJ

As our couples continue to swing into the night I will continue to carry on with my admiration over this lifestyle. Maybe these couples have one up on the average non-swinging couple in the fact that they condone and play together with other partners. This could actually rule out all jealousy by omitting one partner having the affair while the other partner stays home with the kids. If we are really not a monogamist society, and I am beginning to believe we are not, no matter what we were brought up to believe by our parents and church, then playing together may just be the answer to a long and happy life together. At least these couples seem to think so.

I saw a survey on television the other night that had to do with monogamy. According to this survey 80% of the men claimed to be monogamist, and 95% of the women stated they were monogamist. That's all well and good, but by my calculations that leaves 5% of women available for extracurricular play, and 20% of the men playing. If the survey is as accurate as they claimed, then the remaining 5% of those women must be keeping extremely busy.

So, just who is kidding whom here? Are these couples really so happy together? Doesn't any type of jealousy enter into the picture at all? Can they really stay in a happy, fulfilling marriage with the swinging and partying taking place? I have no real answers for these questions, but I would love to know the divorce rate among this group—maybe then I could make a sensible guess at the answers. Perhaps by placing everything in the open leaves no room for the green-eyed monster to appear in a relationship. But, then again if I was a man, I think underneath it all, that I might have a problem seeing my wife with another woman or man—especially on a regular basis. Could be hard on one's masculinity after a bit.

Whatever these couples have that others do not, I found them fun, open and downright enjoyable to be around. They all seemed to have a great sense of humor and really enjoyed talking with others in the chatrooms. There was no animosity or fighting going on with others that is so prevalent in a lot of the chatrooms. I had a really good time with these people, even if I still refuse to take my clothes off in a fully lighted room. ☺

THE TEENAGERS

"Never underestimate the power of stupid people in large groups!" That line came from a profile of a teenage girl who was in a chatroom, and is probably the most profound statement in this entire book. I don't know where it originated, or if she created it herself, but I will definitely make a wall hanging of it for my office.

In this chapter on The Teenagers we have located the good, the bad and the very, very ugly. This chapter proves one does not need any form of human intelligence to bear children, and in certain instances, some children may actually have been born of rocks, although I hate to place this stigma on rocks.

During my travels through the teen chatrooms, I was afforded the luxury (after having had several words with one being who came from a rock) of having this teen say to me, "Hey fat, old lady, get the fuck off the Internet!" "You're too fucking old to be here!" I chose not to take his statements personally as I am sure this twit thinks anyone over the age of thirty is old. While I came across many wonderful teenagers in chatrooms, unfortunately there were too many "pebbles" like this one trying to ruin every conversation. And, ruin conversations, they did.

My questions now center on, "Do any of these pebbles have real parents?" "Do their parents actually teach them the 'It's all about me!' attitude?" "Do their parents teach them how to mutilate their bodies and listen to music that advocates suicide, killing your family, raping your mother or grandmother?" "Do they teach them to be as rude as possible and how to use only four letter words for communicating?" (Did any of you know it's possible to have a complete conversation using only four letter words?) Well, if their parents don't teach them these things then why are these pebbles running amok and annoying the teens, and some of the "old" people on the Internet with their foul mouths?

It seems with most of this pebble crowd that drugs, sex, and (I won't use the term "Rock 'N Roll", for that is my generation) maybe "Screech 'N Zone-out" is the only way to live. Who is supporting these little rocks and all their bad habits? I find it difficult to believe they work for any type of paycheck. Do you suppose the big rocks of the house, who go to work everyday, just hand over money to the little rocks without asking any questions on where they're rolling to tonight, who they're rolling with, and why they are dressed that way trying to disguise what type of rock they really are? Are any of the big rocks giving them the money to buy the music that is advocating killing the big rocks? Nah, I don't think so. I bet all the little rocks have good paying jobs to support all their cravings and habits. But what do I know? I'm just a fat old fucking lady!

Let's go take a look at some of the interesting profiles that will soon be old enough to vote and old enough to run this country. Maybe we can find a future president or congressman in the bunch, or maybe we can find the next Manson or McVey. I don't know who these kids will turn out to be, but I do know one thing, "If the parents of this country don't take their power back from their children and stop being afraid of parental authority, it will only get worse." And, when that happens, I don't want to hear one parent whine, "I just can't control my 12 year old." You see, my grandparents, my parents, and I as a parent, didn't have any problems with controlling our children. We were all raised with one word, and that word was RESPECT. Respect for anyone, anything, and especially respect for our families and life—and this was something they didn't try to teach us at twelve. It was instilled in us from birth.

Here now is what I consider the pebble generation, all found in chatrooms on the Internet.

ID: Female, 15, **LOCATION:** 420 Pothead Ave., Stonerville, MO 42042, **PROFILE:** Marijuana... DUH!!! Hehehehe It's the drugs... im fawked in my head... must smoke more pot... must get more acid... need alcohol... :::DIES::: Im a :::Cough::: Farmer... Theres quite a crop developing in mah closet... "if you gunna smoke with me kid you gunna smoke til you choke." ~RIP~ Matt Cruise

ID: Male, 16, **LOCATION:** Penfield, NY, **PROFILE:** Just tokin' huge bongz, tokin' bongz, drinkin mahself retarded. Be kind 2 kindrz n have a jay=> www.XXXX.com

ID: Male, 16, **LOCATION:** San Antonio, TX, **PROFILE:** I like going to the mall, listening to music like: Slipknot, Mudvayne, Marilyn Manson, Incubus, Hatebreed, Dold Spineshank, Orgy, Soulfly, Nine Inch Nails, Kittie, Disturbed, Coal Chamber, Dope, Staind, and some more. Get this cuz your never going to get me I am the very disease. I work at XXXXXXX (in the gameroom).
I'll always be your shadow
And veil your eyes from states of ain soph aur
I can't be the hero anymore,
I spit up on my plate and then I turn and walk away,
I spit up on my plate and I disrupt the family,
I spit up on my plate and I sever the entity

If you are not sure what a lot of these words are referring to in these kids profiles, I have been told that more than likely they are the rock bands they listen to. It took me awhile to get into "the know" on this genre of music myself.

ID: Male, 17, **LOCTION:** hell hole Texarkana, **PROFILE:** Muzik:::staind, slipknot, korn, godsmack, fearfactory, spineshank, manson, coalchamber, zone, pm5k, system of a down, deftones, a perfect circle, tool, acid bath, orgy, metallica, lords of acid, machine head, linkin park. I love working on my car (66 mustang) and yes I have my tounge pierced. Popularity is a pain that people have yet to discover......I made that up. "u want to play that game bitch," "u make a dash for my cash and it's ur ass that im...blasten"

ID: Male, 15, **LOCATION:** dallas, Richardson, plano, etc...you pick, **PROFILE:** back to old times for real...? Lacrosse, tokin, trippin, drinkin, other sports, and doin all sorts of other illegal shit...wanna know more? Ask me. "it's all about the benjamins, true that be the motto. We ran outta ammo and started throwing bottles..." Is dealing a job? ;\ I'm falling even more in love with you...Letting go of all I've held on to...Standing here until you make me move...I'm hanging by a moment here with you... Life House ~ Hanging By A Moment

ID: Female, 15, **LOCATION:** This huge dent in the middle of California. AAAAMMMMYYYYY!!!!! I have three stalkers that's it. I like scaring little kids, driving down the freeway and staring at people, waving at strangers, being looked at like im a freak cause I am, and drooling over the guys at the hottopic. I sit at home while my parents buy me everything I want cause I'm a spoiled little brat. "you all laugh at me because im different I laugh at you because your all the same", "you are steallar", if you want to destory my sweater, im not insane, one by one the penguins steal my sanity, go wood boy, there all gonna laugh at you, freshmores!! </h7 Never under estimate the power of stupid people in large groups. www.XXXX.com

ID: Male, 16, **PROFILE:** Sunday is gloomy, on shadows I spend it all. My heart and I have decided to end it all. Soon there will be candles and prayers that are sad, I know. Let them not weep, let them know I'm glad to go. Angels have no thought of ever returning you. Would they be angry if I thought of joining you? Death is no dream, for in death I'm caressing you. With the last breath of my soul I'll be blessing you. R.I.P.

A CHATROOM CHAT:

THUG:	you scared to show them to a thug
Nerual:	any female that asks for a hot guy period is a joke
SIM:	hottttttttt
USES:	COMMIN IN HERE TALKING SHIT CAUSE I ANSWERED A Q
Simon:	what
Angel:	ok i can agree with that
SIM:	y are ya so quiet
Nerual:	just a bout 99% of females on internet are jokes!
SIM:	n shittttttttt
Nerual:	they are all superficial!
THUG:	fa sho
Angel:	hey
Bomb:	ISNT THAT SAD NERUAL
Cassie:	No I am not.
USES:	SUPERFICIAL...TRU
Simon:	cause this room is so boring to me
Angel:	roaming in dangerous territory there
Bomb:	ANGEL... YEA RIGHT
Pam:	hey Thug
THUG:	you aint lyin
USES:	N MATERIALISTIC
Nerual:	if you aren't some type of hot guy, they don't want to even talk to you!
SIM:	no it is nottttttttttttttttttt
Angel:	not true
Simon:	yes it is
Linda:	what's happening
Cassie:	Would you talk to a ugly girl?
Pam:	are you really a thug
SIM:	y canyt u just talk 2 me—girl
Bomb:	NERUAL NOT ALL OF US ARE SUCH IDIOTS
Nerual:	why u gettin mad because I am telling the truth
Cassie:	I dont think so
Bomb:	YA KNO NERUAL
Nerual:	looks don't matter 2 me!
THUG:	not a bad one
Cassie:	Ya right

Nerual:	it matters on someone's personality not how they look
THUG:	i just be chilln
USES:	LOOKS MEAN ALOT TO ME
Simon:	you want me to talk to you
Bomb:	LOL USES...
Angel:	sorry but i can say the same for all you guys with your where 's the pictures of girls ass
USES:	IF U DONT TURN ME ON
Cassie:	attraction is everything
Pam:	oh what kind are you then
SIM:	yup
Rae:	u people say girls are jokes did u guys ever think it might just be u
USES:	WHATS THE USE???
Bomb:	ANGEL ITS AN ASS...
Angel:	i agree
SIM:	yesssssssssssss
Nerual:	I don't look good and I don't care if I talk to some fat busted up girl because she probably
Party:	the name of this room is a joke
THUG:	im just a straight up ass dude
Pam:	oh ok
Bomb:	TAHTS COOL THUG
THUG:	ya feel me
Bomb:	AT LEAST UR HONEST BOUT IT
Linda:	i aint madd i am just sayin dont disrespect us females cuz yall are on lookin for hotgirls
Party:	everyone knows that males are jokes
Nerual:	is a lot prettier then most of these "hot" girls on here!
Pam:	are you putting it down for anything
USES:	I AINT HOT...BUT I AINT BUSTED NEITHER
Simon:	yr so full of it
Bomb:	PARTY OBVIOUSLY NOT ME
Bomb:	CUZ I DISAGREE
THUG:	what u mean
Rae:	hell yeah party
USES:	WELLLL...ATLEAST I DONT THINK IM BUSTED
SIM:	i am—y do u have 2 say it like that—girl

Angel:	like i said Nuetral i could say the same about guys or worse
Pam:	are you in anything
Nerual:	looks really should not matter!
Nerual:	guyz are even worse!
THUG:	no inot in a gang
Nancy:	BUT U KNOW MOST OF THE HOT GIRLS R JOKES ANY WAY
Bomb:	AND THATS ALL THAT MATTERS USES

ID: Female, 16, **LOCATION:** Loserville, **PROFILE:** Picking on old bold men with hairy arms…Throwing Monkey Turdz at preps…Talking to my friends and others…I like walking up and down the street and watch religious girls suck face and bless little children with holy water. A True Loser/Wicca/Crazy Elf that excaped from the North pole!!!If u look down u might just see me sitting under your chair. "I hope life isn't a big joke…because I don't get it" & "I smile because I have no idea whats going on" Love is like a violin. The music may stop now and then, but the strings remain forever." ~BYE~ www.XXXX.com

ID: Male, 17, **LOCATION:** Hell, AKA Mormonville, USA, AKA Ogden, Utah, **PROFILE:** you laugh at me because I'm different, I laught at you because your all the same-Jon Davis. It's better burn out then to fade away-Kurt Cobain. I'm a Part time @ss hole, Full time d^ck. I'm not like them but I can pretend theday is done im having fun I think im dumb maybe just happy think im just happy think im just happy think im just happy my heart is broke I have some glue-kurt Cobainn

ID: Female, 16, **LOCATION:** West Coast, **PROFILE:** I'm a bitter young woman, Don't Piss Me Off=X. Music, Reading, Writing, I'm a student. My motto: Want. Take. Have.

ID: Male, 16, **PROFILE:** This generation they won't keep quiet. Work Work Work or Riot…BUSINESS Sex& Violence…The Exploited Camel City Punks and Skins Nc. Its your hate on which we feed. We are the new class we are the new breed. Send our regards to a nation on fire, and with love a bouquet of barbed wire—Blitz. Street Punk, Oil, 80's New Wave, Yelling and Sleeping. Its another boring night, and I'm feeling pissed. My heads f*cked up and I'm in a mess. To many drugs they make my high, I wanna cause havok I wanna die…-Anti-Nowhere League I got a Sunday morning nightmare-SHAM 69 Hang'em High-Combat 84 Warrior…Don't pass go. Don't post bail. Rich people don't go to jail!…Anti-Heros Get Pissed! Destroy!…-Sex Pistols I'm an upstart! HEY! What are you gonna do!? I'm an upstart! LISTEN! I'm talking to you!…-

Angelic Upstarts Bullsh*t Crass! You've been detected!...-Special Duties
www.XXXX.com

ID: Female, **LOCATION:** CA, U.S. of Oi!...santa Fl_lCKIN' boring cruz, **PROFILE:** We are 138-androids are people too, right? "my baby got ran over waa waa ooo by a steamroller" beer and sex and chips and gravy, that's all that a maccLASS wants!   I'm Ronald McRegan, McDeath, McNuclear, McWar...McTommys McFear me, cos I'm McDangerously Crazy!" GBH, Stiff Little Fingers, Menace, Vibrators, Crux. The Adicts, Major Accident, Templars, Blitz, Anti-Nowhere League, Sham 69, Vice Squad, Oxymoron, Unseen, The Enemy Partisans, Dayglow Abortions, Misfits, Angelic Upsarts, MaccLads, Last Resort, StiffLittle Fingers Special Duties, Suicidal Supermarket Trollies, Peter and the Test Tube Babies, Cock Sparrer Slaughter and the Dogs WORK IS THE CURSE OF THE DRINKING CLASS...Fl_luck! "Where have all the boot boys gone?" ...I'll die with my boots on!

It is interesting how the black teenager can use words that berate their own race. I was told these words, when used by a black to another black, is a sign of affection. When used by a white to a black it's a sign the war has just started.

ID: Male, 17, **LOCATION:** IM REPING DAT DIRTY SOUTH....SCREAMING ATL FO LIFE, **PROFILE:** HANGING WITH DA NI99AS AND MY GIRLS YA NO!!!!!? TAKING YALL PLAYA HATING NI99AS GIRL... HA HA HA IZ ALL ON YO PLAYBOY!!!!!!

ID: Male, 18, **LOCATION:** NYC, **PROFILE:** Hold Me Down Like Woah!—Rippin Up Dese Streetz in NYC—NYC Clubs-Imagine Friday Nights @ Exit Its where im at- Sex: Bring It On!! Sports, Chillin, Parties, Bein my Badass self, Parties, Girls, Driving the NY Machine!!! Listening to music like 3 Doors Down, Papa Roach, Mystikal, Red Hot chili Peppers, Goo Goo Dolls, OutKast, Metallica and much more!Im not a tease, im just a reminder of what u cant have0-We were born sweet little babies but the New York streetz raised us crazy!-Im not a bad person I just like 2 do bad things-Representin da city that never sleeps n the parties all night long...NYC!!!

A CHATROOM CHAT:

Rage:	<— 16 M Pic
LoveU:	"you shouldnt do that to yourself
ICP:	rage, ic me ;0)
RED:	HARD CORE
LoveU:	then i ran my foot up there asses

Rage:	uhh ok
KILLER:	sent
Vive:	punk iz bad ass...just not new school...
Death:	yup
Death:	old school is the only punk
Vive:	are you agrreing w/me death...
Princess:	15/f/ga/cherry red hair/eyes change colors/5'6/freak/im me 2 chat
Andy:	hehe… watch out,... my Erin is dead sexi... it may cause your brain to explode with
Andy:	dirty thoughs
Death:	yeah i am
Vive:	well dont...
Andy:	*thoughts
RED:	SEXSEXSEXSEXSEXSEXSEXSEXSEX
Death:	well go fuck yourself
KILLER:	ahhhh lol
Krack:	anybody like goth freaks
KILLER:	no i'm not
Death:	my opinion stays the same
Vive:	iggy and the stooges was good...not iggy by himself…
KILLER:	goddamn you guys
Rage:	i have very little thoughts... but when i do half a thought, half of them are dirty and half
Rage:	are illegal
ICP:	i know what's on red devil's mind ;0)
RED:	SEX
Vive:	dammit death...who are you…
Rage:	<— fucking is on my mind
RED:	PENIS
ICP:	BINGO!!
Grand:	thinking is amazingly groovy
LoveU:	any bitches that l9ike guys with peircings and tatts
Brat:	hey peeps
Death:	hey killer...they picking on you?
KILLER:	shut up penis wrinkle
KILLER:	i don't think so
Andy:	u call ME penis wrinkle??
Andy:	=(
Death:	why not...you suck

KILLER:	no!
Rage:	i am glad i dont have any wrinkles in my penis
Death:	haha
LoveU:	suck
Vive:	its spelled "JUGGALOE*
RED:	PENIS IS GOOD
Brat:	16/f/fl any one wanna chat press4:20 or ic me
KILLER:	i was talking to red
Rage:	nu uh
LoveU:	i have a penis
KILLER:	so what
Cradle:	16/gothic chk-freak/fl/pix n profile
ICP:	yeah, i know but you can't put it on there cuz somebody already had it stupid
Andy:	ok... whew
ICP:	god, what's your profile
RED:	I HAVE A PUSSY
Vive:	BUT YOU WOULNT KNOW THAT SINCE YOURE NOT FRON DETROIT...
KILLER:	goddammit...
Death:	haha...upload's done
Death:	alright
Smirk:	i have a big dick
ICP:	ugh, i know
KILLER:	shut up red
ICP:	lay off psycho
KILLER:	::licks alice::
LoveU:	i only have a small 8 inches
RED:	I HAVE BIG BREASTS
Rage:	I like boobies
KILLER:	big tits suck
LoveU:	i'm thinking of spending 2 grand to add 2 inches
KILLER:	they are highly over rated
Satfaction:	mine is 8.5

ID: Male, 17, **LOCATION:** IN HELL, **PROFILE:** WHAT DO YOU CARE. I AM A LAZY BUM. NO JOB IM A LAZY BUM REMEMBER. GO TO HELL WHERE YOULL MEET ME!

ID: Male, **LOCATION:** reppin Da heartbeat til I RIP. **PROFILE:** Stay gangsta Keep it thoro til da end...i dont care who u wit or who u get or what u

146

got all of dat will get u shot everybody in da world kno—is hot. What I look like turnin' down chocha a hoe gone be a hoe n a nigg a gone be a nigga

Many teenage profiles proclaimed Marilyn Manson as someone to be respected and admired. The rest of us just don't get it. Well, they got that right! I don't get it!

ID: Male, **LOCATION:** TEXAS, **PROFILE:** Student and oh whats that job I do again? Nothing…a worker for LCF and AT. Is adult entertainment killing our kids or is killing our kids adult entertainment? M. Manson. He had nothing 2 do w/Columbine so shut the hell up! Look where its at Middle America…Now it's a tragedy now its so sad to see…no one cares anywhere else…M.

ID: Male, 15, **LOCATION:** The pretty place where the flowers grow…I'll be back in an hour or so, **PROFILE:** Like I am gonna say my name! I don't want some freak comin' up and saying "Hey—" Bladin, Boardin, Snowboardin, surf'in, KickBoxin I like x-treme sports. A ghetto a*s piece of sh*t!!! Student!! HELP!!!! F*ck All You! If life throws you a grenade pull the pin and throw it back! Lifes a big fuc*in game so play it…And Cheat!!

A CHATROOM CHAT:

Strawberry:	sure ur 14
Strawberry:	I bet u guys are like 12
Hottie:	123
Alx:	14/m
Shortie:	bh…asl
Dragon:	Who Strawbery?
Gods:	ANY BOYZ WITH A BIG DICK PRESS 978
Hottie:	14/f/cali
Bighen:	who in here is over 18
Strawberry:	all the 14 year olds
Bighen:	978.com
Shortie:	ic me if you wanna chat hottie...
Strawberry:	I bet there all 12
Dragon:	Ummmmmm why?
Shy Girl:	i might be bighen
ZuluBomb:	everyone in here is probably like 11 and 12 and pretending that they are 14
Strawberry:	because they all LIE
Strawberry:	yup
Avis:	im really 14

ZuluBomb:	u are all liers!
Strawberry:	and they all rape there hampsters
ZuluBomb:	muah ha ha!!!!!!!!!!!!!
Dragon:	Now Strawbery how old are YOU
Strawberry:	LOL
Bighen:	TRIXS ARE FOR KIDDDDDSSS
Strawberry:	72
ZuluBomb:	ha ha
Dragon:	And Zulu, can you talk for yourself?
Strawberry:	yup
Shortie:	no no i wanna chat wit boys any 13 or 14 males wanna chat wit a single gurl ic me~or123
ZuluBomb:	not really im 16
Man:	<~ ~ ~ ~ 17/ M/ CALI/ PIC S2R ONLY/ IC ME TO CHAT
Gods:	ANY BOYZ
Strawberry:	lets sing the frosted flakes commercial
Avis:	14/m
Hottie:	14/f
Shy Girl:	man what city
Dragon:	15-Female-New York-Pics
Gods:	13/F
Strawberry:	I like the things you do
Kitten:	hey room…
Man:	LA
Bighen:	philly
Avis:	14 / m / ny
Strawberry:	I wish I could be you
Hottie:	im really 13 peps
Kitten:	16*f*iowa*pix
Bighen:	why
Man:	U??
Strawberry:	Ur the 1 and only tiger with
Strawberry:	the 1 and only taste
MATT:	n e girls wanna chat wit a 15/m im me
Strawberry:	frosted flakes there more than good
Man:	<~ ~ ~ ~ 17/ M/ CALI/ PIC S2R ONLY/ IC ME TO CHAT
MATT:	n e girls wanna chat wit a 15/m im me
Strawberry:	there GREAT
Gods:	I WANT TO CHAT
Strawberry:	I wanna talk

Strawberry:	to someone
ZuluBomb:	i am so fucking bored

ID: Female, **LOCATION:** some place I wish I wasn't, **PROFILE:** I am the Electric Messiah* <(Reach out and touch faith)> you can turn your back on a person, but NEVER turn your back on a drug specialy when it's waving a razor sharp hunting knife in your eye. Husband&kid(inthefreezer)yum With Satan You Just Tell Him What You Want And You Get It—EHC Blah: For Beauty Is Always Cruel ****everything that give pleasure is evil****Confess your sins to Me Z

Here I thought I had found a young person who was actually going to say something worthwhile and give me hope. Read on.

ID: Male, **PROFILE:** yo I wrote a poem a long time ago/a man I use to know told it to me/it went something like this/and let us cease with this anger and hatred toward men/from men were born to brothers, thru friendship not war/and let us not seek to find friendship with handshake nor with promise/for peace must be nurtured until the very thought of violence is been removed from our eyes (this is me responding to the mans poem)...man forget that shit/the world has been broken into billions of pieces/thus givin me, and everyone like me from the ghetto/nothing, cuz all were doin is strugglin for a peaceful event/you gotta come stronger than that g/man where im from is a modern day Vietnam/and my backyard is concentration camp/its not all about bein militant/its all about survin man/dah hell with the government/cuz all the government know how to do is not govern.

A CHATROOM CHAT:

Soccer:	hott 13/f/here press123 to chat
Crystal:	hi room
Kassie:	sup guys
Kassie:	all guys age please
Ball:	hey room
Ball:	n e 1 wanna chat with a 13/f/il/pic press 000 or ic me
Crystal:	asl every-1
Soccer:	13/f/cali
Crow:	play you know my name
ImSking:	000
Shadow:	13/f/tx press 103 if u r under 15
Crystal:	any 1 want to talk

ImSking:	103
Playboy:	Was up everyone 16/m here want 2 talk
Crystal:	hello??????/?????/?????
DARK:	WAZZ UP ROOM
Crow:	18/m/mi/pic
DARK:	A/S/L EVERYONE
Crow:	stop
Crow:	Eric Draven< støpped: [south park mexican - you know my name]
Crystal:	103
DARK:	13/M/OR

The below profile stated he was eighteen years old, yet I found him in the above chatroom with thirteen year olds. If the occupants of that chatroom are really thirteen years old, I would like to know what this eighteen year old could possibly have in common with them, and vice versa.

ID: Male, 18, **LOCATION:** Kalamazoo, MI USA, **Profile:** Not married yet, drinking till I die, student, you didn't know the boogie man was a clown, but when you see the juggla, your holdin your juggula

ID: Male, **LOCATION:** The Bronx, **PROFILE:** WHATEVER I FUXIN FEEL LIKE DOIN WHEN I WAKE UP…I DO… PEACE TO ALL MY HOMIES. I LOVE U ALL MY BROTHERS…LIVE UR LIFE LIKE IT"S 1 MORE ROAD TO CROSS…CLICK CLICK CLACK UR SHOT IN UR BACK…LIFE IS LIKE A BOX OF CANDY U NEVER KNOW WHAT UR GANNA GET…REST IN PEACE BIOTCHZ!!!!

This young lady sounds like someone who knows everything there is to know about life. She sums it up in one sentence. "If you ain't real, you ain't shit!" Perfect statement. I couldn't have said it better! Who could possibly need further education with that intelligence going for them?

ID: Female, 18, **LOCATION:** H-Town Representa Soufeast 4 Life, **PROFILE:** (There can only be 1 me) Chillin and Staying on my Grind 4 dat 01- If you Ain't real You ain't Shit!! I love writing poetry so if you want a sample let me know. Fu*k what you heard, Shake dem hataz off! Lights off unplug the phone I'll make him eat it while my period, a lil nasty yo, Red Bone but I'm classy though, ya ain't ready. To all ya'll freaks, don't ic with that bullshyt azz question"R u a Freak?" www.XXXX.com

A CHATROOM CHAT:

KING:	<~~18/m/puerto rican/bklyn (E.N.Y) with pic any females wanna chat press 731
ANGEL:	ITZ ALMOST MY B-DAY
Shelly:	i thought u was my age
ANGEL:	ILL BE 19 THA 15TH
ANGEL:	WOOOHOOO
KING:	<~~18/m/puerto rican/bklyn (E.N.Y) with pic any females wanna chat press 731
Baller:	ladies wit pics 2 trade hit 6969
Shelly:	HEy
Piece:	SHELLY U CAN IC ME
Not Famous:	i need a ride or die bitc*
ANGEL:	RIDE OR DIE BITCH RIGHT
ANGEL:	<<<<<DIRTY SOUTH RYDE OR DYE BITCH FA SHO
Not Famous:	where my crips at
ANGEL:	SORRY
ANGEL:	NO CRIP HEA
Shelly:	LoL
Shelly:	No crips?
Shelly:	i have pixs
Shelly:	but im a pennie piec
Not Famous:	any locstasin here
ANGEL:	WHERE MY BLOODZ
BaBy:	any fine ass horny blackguys wit pics ic me
Piece:	NO BLODS HEA WHODDY, ALL BOUT DAT 60 CRIP...U HEARD
Piece:	LMAO
Not Famous:	CRIP 4 LIFE
Playa:	IM COLOR BLIND
Playa:	FUKK BLU & RED
Lingtin:	i am purple
Not Famous:	FUKK NU
Piece:	YEA PLAYA IZ ALL ABOUT THAT PURPLE
HotGirl:	PLAYA CLAIM DOWN
Not Famous:	6 POPPIN 5 DROPPIN
LilThug:	i just smoked a blunt to my dome and got laid i feel great right now

ANGEL:	WELL YA BETTA GET YA FUCKIN COLOR BLINDNESS FIXED AND SEE WHAT COLOR YA BLEED
ANGEL:	5 POPPIN 6 DROPPIN
ANGEL:	SEE IT TAKE 6 YALL TO DROP 5
Shelly:	im colorblind
ANGEL:	IT TAKE 5 US TO DROP 6
Shelly:	thug u know how dum u sound?
LADIEWET:	17/F/IL
Not Famous:	I BLEED BLUE
ANGEL:	I BLEED RED
Shelly:	blood is blue...
Shelly:	in ur veins
Shelly:	SO HA
Playa:	MY HEART PUMP BATTLE FLUID
Not Famous:	SPET YOUR SET
ANGEL:	NAW ITZ ONLY THAT AWAY TILL IT HIT AIR
Playa:	NIGGAZ BEEFIN OVA COLORZ
Playa:	MAAN...
ANGEL:	BUT WHEN YA SEE IT IT RED
Playa:	WHERE YALL SETS AT
ANGEL:	AND YOU CANT SEE IT WHEN IT BLUE
Juicy:	<<<<Maryland nigga with pics
Playa:	IN A ROOM SOMEWHERE
ANGEL:	BUT IM GONNA JET
ANGEL:	~BLOOD LUV
ANGEL:	B` UP
Shelly:	OK WHOOOOOOOO cares
Not Famous:	U PUS
PERU:	WERE R THEM FINE NIGGAS N LATINOZ AT?WANNA CHAT? IC ME
Shelly:	U dum
Piece:	DAMN MUTHAFUCKAZ...YALL NEED 2 QUIT WIT THAT OL SHYT...GROW DA FUCK UP

ID: Female, **LOCATION:** Montgomery Alabama, **PROFILE:** Still in high school, single he wasn't man enough for me. I be too hot too handle so if u touch u might get burn, so u got to be a Tru hotboy to get with this hot girl.

ID: Female, **LOCATION:** Fort Myers, FL, **PROFILE:** *~*80% angel 120% Devil*~* LiL Colombian Princess*~*Naughty Flirt*~* ~-I::*BIG Pimpin up in SCP*::I-~ Why??? Wanna come ova?? Want my phone # too???? **1-800-U-B-XXXXXX!!!** Guys <3, chillin at peepz crbz, shopping, watching moview with mah crew...Just hang out and chill with mah homiez and sistaz... I LOVE YA!! AHOUT outz to ya'll—There is just not enough room for everyone!! —**DDD** ~~Cross Country and Track Runner~~ and a SENIOR AT Estero High c/o 2001!!! Don't drink and Drive... U might hit a bump and spill ur drink!!!! $ex is like Pringles...Once u pop u cant stop!!!!! :-i::* I'LL be all yoo eva need, satisfaction guaranted*::i-:

ID: Female, **PROFILE:** smiles OK...WHO DID IT...WHO JUST GRABBED MY AZZ? If I can touch your ass, then I will you drive my car K? I didn't touch your ass cause then I'd get addicted. WELL I WANT TO TOUCH UR ASS DAMMIT. I WANNA FUQQ. If I touch ur ass...will u scream my name? Touching ur azz mekes me ug...::drip drip::uh un oh I gotta get new panties 8)~ **touches ur azzz** MMMM is that a nice booty or is it jus me? OH MAY I? ::touch::OOOH::creams::

ID: Male, **PROFILE:** *Ticalion Stallion* if u know what I mean ;) DA WooDzz WHA! Baby don't cry Gotta keep ya head UP even when the road iz hard neva give up. U aint hype to die...but u hype to shoot. Hey what can I say I'm just a thug nikka. Play wit ma white chocolate salty ballz *PLEAZE!* ~~>PC DC Ma Thug NikkaZz<~~

ID: Female, 14, **LOCTION:** New Port Richey...Florida...near Tampa, **PROFILE:** ~~~Single~~~ and lovin it! Chillin with my homegurlz and my Papi's, partying, booty dancing. Stundent. If my back and hipz don't hurt in tha morning...U didn't do the job right! Don't Playa Hate...participate! If you can make me Cream...I can make you Scream.

ID: Female, **LOCATION:** NYC...Money Makin' Manhattan (Where da city neva' sleeps), **PROFILE:** ***I'm a 100% LESBIAN therefore No Guys & No Cybering*** **Just tryin' to make friends** Driving around, blubbing (sometimes), movies, biking, blading, pool, bowling, listen to music (all types)...da basic fun stuff but also like to be relaxed & laid back & just chill...I like to make people laugh & keep them smiling! I'm good at it too! More about me: I'm a romantic soft soft butch in search of Bi/Lez friends...think ju can handle dat?? =o) Full Time LaGuardia Student & getting' my $$$$$...bet you wish you knew how huh? Well, that shouldn't really matter to you! Just be real and we'll do fine...men DON'T waste your time (or minez) tryin' to get to know me cuz I'm STRICTLY for da ladies!!! ~*AiGhT!*~

The following chatroom is called an insult room. Teens can go there and dispense insults at the other occupants in the room. What an interesting way to spend the evening. I can't think of anything more delightful than being on the receiving end of some of these remarks.

A CHATROOM CHAT:

Tyler:	bren u geek shut the fuck up u havent said shit funny
Endig:	Jenn <—knows eating oats, It's a fat sows basic diet
Jenn:	Endig<—steals hubcaps with her mama, cant find her daddy. He is shining shoes somewhere
Endig:	Jenn <—spends her time on the beaches, dodging harpoons
Quacks:	no i was looking at the pic of ur mom that i keep on the back of my toilet
Bren:	tyler if i wanted any lip from you i would scrape your teeth off of quacker zipper
Jenn:	Endig<—spends her time in the ghetto dodging bill collectors
Endig:	Jenn <—poster model for "why you should not breed your dog"
Jenn:	Endig<—poster model for slavery
Key:	CLICKED THE CUNTIEISHER
Gunman:	who in here worships satan?
Endig:	Jenn <—has no bills, how much does a Kmart box cost??
Tyler:	bren if i wanted to hear u id un stick your tongue from my dick
Jenn:	Endig<—only has money first of the month like the other sistas
Endig:	Jenn <—knows slavery, think that sums up her life in general
Quacks:	lol tyler
Monkey:	bout damn time lol
Jenn:	Endig<—slavery sums up her heritage
Monkey:	fuckers
Endig:	Jenn, maybe give up prostitution and get a real career
Tyler:	sup street

Cutie:	whoa
Endig:	fat lady in the circus??
Monkey:	sup
Jenn:	Endig—and maybe you should give up welfare
Bren:	tyler you should know from exspearance
Cutie:	no you are a fucker
Cutie:	not me
MCee:	fuck u all
Tyler:	u should know from english class
BUBBLES:	endig~ and cutting hair is a rich career?
Monkey:	ur mom is 1 too
Cutie:	and don't copy my color, you biter
Bren:	wut happen quacks have you stiuck on his cock
Tyler:	u failed it right u dumb fuck?
Joey:	÷^±/ Ràmþàgè †øølz ²·º ß Øøglè
Joey:	÷^±/ Úñlöådèd bý Jœý°³³°
Cutie:	i'm talking to u monkey
Quacks:	breni told u u needed phonics
Cat:	whoa...spell check, bren!!
Endig:	Bubbles you don't have hair, so why worry
Jenn:	oh and thanks Tyler for the IC before
Tyler:	exspearance?
Tyler:	its cool jenn
Quacks:	tastes better than urs
Lisa:	REALLY CAT LOL
Endig:	Bubbles soon as you reach puberty, get back with me K?
Cutie:	probably u named yourself a monkey cause u smell like one
BUBBLES:	U been looking at my puss????
Jenn:	I was too busy to IC back, but thanks
Monkey:	u talking to me? good meet ignore
Monkey:	lol
MCee:	Loaded Hacker with Methodus Toolz
MCee:	Hacking: Joey
MCee:	Requesting Joey's IP
Gunman:	fuck you
MCee:	Checking File/Print Sharing ports
MCee:	Open port found
MCee:	Brute Forcing c:\ sharing password
BUBBLES:	i'm 18 asshole
MCee:	Password Found

Endig:	Jenn <—reason why Mcdonalds has sold a Billion Hamburgers
MCee:	Deleting important system files
Key:	HEY ENDIG...GET BACK ON YOUR STROLL BEFORE DUANE BEATS U AGAIN
MCee:	Process Completed
Endig:	1 billion she served, herself
Cutie:	you talking to me??/if you are shut the hell up
Cutie:	gosh
Cutie:	u
Cutie:	guys
Jenn:	Endig<—reason why afro sheen has sold a million spray cans
Cutie:	are
Cutie:	so
Cutie:	boring
Tyler:	hey bren where u at dumb fuck?
Endig:	Key go lay down, before i get out the 220 volt electric shock collar and zap you inot
Endig:	into next week
Jenn:	Endig is as black as I am fat
Quacks:	hes jealous of u
Monkey:	clicked cutie for being a dumb ass scroller
Quacks:	tyler
Lisa:	HE IS GETTING A DICTIONARY TYLER
Endig:	Jenn, nothing wrong with being black, Im sure it sums up your behind
Key:	YOU'D NEED HELP PLUGGIN IT IN YOU STUPID FUCK
Tyler:	why cause i can spell?
BUBBLES:	how old are U endig?
Endig:	maybe try a bath once in a while??
Jenn:	Endig—dont make me put on a pillow case with eye holes and burn a cross in your yard
Endig:	Bubbles? why you need a mommy??
Quacks:	i said that u taste better
Jenn:	Endig—best way to get you out of the water is to throw in a bar of soap
Gunman:	look up chode
Endig:	Jenn? Im sure if anyone is burning anything in your yard, most likely it is
Cutie:	shut up monkey

Drew8863:	YOU INSOMNIAC!!!!!!!!1
Key:	ENDIG...PUT DOWN THE CRACKPIPE AND PICK UP A BAR OF SOAP ONCE IN A WHILE
Jenn:	Endig—you type very slowly, dont you
Endig:	to rid themselves of the crabs you lent them during sex
Monkey:	all others fuckin newbies lol
Jenn:	I got those from your tongue
Cutie:	wuever
Monkey:	wuever?
BUBBLES:	endig~what are U 80 years old and on the internet????
Quacks:	what happen bren get lost in ur moms ass
Endig:	Key? you know soap bending, this is not the time to get into your prison adventures K?
Drew8863:	LOL YOU KNOW DECIET RIGHT????????
Monkey:	illitarate fuck
Cutie:	wutever!!!
Cutie:	don't you noe how to read!
Monkey:	noe?
Endig:	Bubbles if I were 80, bet i would be slow enough for you to catch
Monkey:	dumb ass kid
Endig:	blind enough and weak enough to not really care that much??
Jenn:	Endig—you are slow enough because you are fat
Bren:	quaks your so bright for someone who`s gay and like`s to get his stool`s pushed
Quacks:	sorry i m a she not a he
MCee:	Loaded Hacker with Methodus Toolz
Jack:	HEY PEEPS I JUST CRAPPED REAL BAD
MCee:	Hacking: Monkey
MCee:	Requesting Monkey's IP
Tyler:	bren used spell check thats what took so long
MCee:	Checking File/Print Sharing ports
Endig:	Jenn, go shove another donut in your mouth maybe you will attract some police officer
MCee:	Open port found
MCee:	Brute Forcing c:\ sharing password
MCee:	Password Found
MCee:	Deleting important system files

MCee:	Process Completed
BUBBLES:	endig needs to grow up and get a piece of ass
Jenn:	Endig - collect yo food stamps fo you illigetimate chilluns.
Endig:	Jenn <—"Im skinny = First indication that she does not look good in spandex
Jack:	IT HAD BEANS AND CORN
Quacks:	accually he just got his head unstuck from his moms ass
Endig:	Jenn <—knows welfare lines, it makes up 99% of her income
Jenn:	Endig<—says I am fat—first indication she has no other ability at insult game
Tyler:	haha
Lisa:	WHY END DO YOU HAVE TO GO ON ENDLESSLY ABOUT JENS WEIGHT
Endig:	the other 1% comes from selling crack and her body
Lisa:	TAKE CARE OF YOUR PROBLEM
Jenn:	Endig I had out more in tips than you get in welfare
Quacks:	he had to call the plummer
Monkey:	<—hooked on Jolt lol
Endig:	Jenn <—if selling her body were her first pastime, she would surely starve
Jenn:	I am 100 lbs and only 5ft tall, but she doesnt know anything except fat one-liners
Tyler:	endig wanted her pic on here but they havent mad 35inch monitors for her fat ass to fit on y
Key:	JENN ISNT THAT FAT...ITS JUST THAT DOORWAYS ARE SO SMALL

ID: Female, **LOCATION:** Las Vegas, **PROFILE:** ~Snowboarding*14 piercings* concerts* parties* etc~ Party hardy, Drink Bacardi, Rock'n'roll, Smoke a bowl, Do a line screw a few. Were the class of 2002* Student at Centennial. Be yourself don't act like someone your not...or youll never find your true self!

I haven't said too much during our travel with the teenagers, as I felt my words were not needed, and what could I possibly add to what you have just read?

Maybe my next book should cover the terrific teens with their wonderful outlooks and sense of humors, for I believe one would need an incredible sense of humor to make it through the teen years today with peers like some we have just experienced. Hopefully, most of the teenage profiles here are nothing more than wishful thinking, but even so, it is scary just to know this is where their minds are at at this moment in time.

If you have found your child here, now might be a good time to discuss what is happening in their life with them. Whatever the outcome of these children, they all belong to someone and I believe that someone cares deeply for them. Please, let's not allow our children to be throw-a-ways.

CONCLUSION

Our adventure exploring the world of the adult Internet chatrooms has come to an end, and I hope it was as great an adventure for you as it was for me. It was certainly an eye-opener for me, as I was aware such lifestyles existed, I just never knew those that practiced them were so open about it, and I thank them for letting me into their world—if only for a brief second.

I have been given an education in Life 101 and can't wait to begin Life 102. I have made new friends and learned a lot about other people and their preferences for what they would consider a happy life, and I learned a lot about myself, even at this late stage in the game. My people watching has been expanded as I now wonder what that couple sitting in the corner of the restaurant are really doing when the sun goes down, as with the couple on the beach, or those two men shopping together in the market. With so many people in the world, how many are actually as straight in their sex lives as those magazine surveys would have us believe? There are millions of people on the Internet, and while I admit not everyone is swinging from the chandeliers, my research says a lot are. In fact, more than I would have ever imagined.

It appears men and women, no matter how they would like to deny it, have this other side to their outward appearance. Maybe it is nothing more than a fantasy side, but as the saying goes, "If you dream it long enough you will eventually create it". So I am safe in stating, "Not everyone uses the missionary position!" And, probably 95% of you reading this are now saying, "I am monogamist." "I am straight." "I have never done these things." "Why I have never even thought of them!"

"Hmmmm, never?"

The people that are engaged in the lifestyles that we visited all have one thing in common. They truly like the lifestyle they are living. I found no one wanting out of a relationship because of the way they were being treated sexually. Now, if one were to ask the question, "How is your sexual relationship with your husband?" to a group of monogamist women, say married for at least 4 years, the answer would be a lot different. The majority of women and men in the average "vanilla" relationship are unhappy. Don't believe me? Then how do you account for all the sex books, magazine articles on sex, and sex therapists? Millions of dollars are made every year off of people looking to put spice back into their life, or trying to find new ways to keep a marriage alive in the bedroom, or just trying to get a husband or wife to notice them again. Maybe if couples could just tell each other what their fantasy is, without being judged, a lot more would stay together and really get to know each other. After all, some of your husbands and wives are probably right here in this book, in a chatroom, being someone else. Look what you might be missing in your own bedroom.

Since we have only explored the "sex" chatrooms I feel I must give due to the other chatrooms available on the Internet. There are some terrific chatrooms available on a daily basis in which you can make life long friends. I stated at the opening of this book that I met my husband in a chatroom, but what I didn't mention was the fact that I was lucky enough to find a chatroom in my early days on the Internet and that the occupants of that chatroom actually saved my life. Simply put, I had had emergency open heart surgery after a medical procedure went wrong, and when they released me from the hospital I was so beat-up and depressed that I didn't know how I was going to make it. I couldn't open a door, a cupboard, or raise my arms. What a mess I was. Then I found this chatroom with these incredible people who were having so much fun just talking, telling jokes, and laughing. I started to visit this chatroom regularly and before I knew it I was actually laughing again. And, these people cared. They really cared about each other. Since that time I have become close friends with several of them, and flown out of town to attend weekend bashes. So you see, I owe the Internet and chatrooms a lot, as they literally gave me back my life, laughter, and of course incredible shopping.

You can find chatrooms for people in their 20's, 30's, 40's, 50's, even their 60's and 70's. There are chatrooms for almost every type of hobby, entertainment, or friendship you could think of. If anyone is sitting home and feeling lonely, all they need to do is get on the Internet and go exploring. It may take a little time to find the right group to click with, but once you do I guarantee you will not be sorry. There really is life out there even if you can't leave your home, and I cannot possibly imagine what my life would have been like without "my" chatroom.

While I may not agree with everything that I have chosen to use in this book, and while I do not personally believe some of these people should be allowed on the Internet, I am in no way condoning censorship. I do not believe in censorship in any form for if we start to censor one thing, then we must continue to censor another, and no one has a right to tell someone else what he/she may read or do if it does not violate any laws or affect one personally. It is called freedom of choice, and the Internet should remain an uncensored entity remaining a free choice for all. If you don't like the content, then don't go there, change the channel, or close the book. You can also take the time to teach your children and become a part of their lives instead of turning them loose on the Internet to discover life from some of the chatroom occupants that you have just met, while you are so busy doing whatever it is you do. Let your children know that people really count in this lifetime and that life and caring is worth more than money can buy. And, if your teenagers are so damn bored go buy a dairy farm…everyone knows a kid can't be bored when they have to get up at 3:00a.m. to go milk cows. I never heard of a teenager on a farm, who helped bring in the crops, tend the livestock, and run the machinery saying, "I'm bored, and there's nothing to do."

The chatroom chats that I have chosen to use in this book do not reflect the kind of chats that take place in your everyday common chatroom. While unwanted persons can infiltrate chatrooms of any type, those persons are quickly reported and not tolerated by the regular room occupants. So you can lay most of your fears aside that you will not run into these chats unless you specifically go looking for them. What you will find, for the most part, is good clean fun. When a chatroom is disrupted by a person who wants all the attention, uses foul language, and in general shows no type of human intelligence, it is usually a teenager being bored with life. Can I mention the "cows" again?

Feeling that I have said all that I can on the subject of chatrooms, and hoping you have enjoyed your journey with me, there is one more thing to say. If most of the profiles we have read are nothing more than an active imagination, then I would like to leave you with one more active imagination and what I would consider a perfect profile.

NAME: Kitten, **CITY/STATE:** Her Apartment, **STATS:** 23 yrs. old, 5'5", 115 lbs, 36-24-25, brown hair, blue eyes, **DESCRIPTION:** The aroma from the kitchen is infiltrating the apartment as she lifts the lid from the pot on the stove. The boiling water swirls as she reaches for the lobsters showing them their final resting place. She turns as the doorbell rings, and glances in the mirror. Her blouse is perfect—deep crimson red, low cut, silky and clingy. It outlines her breasts in just the right manner. Not too little. Not too much. The skirt is also perfect. Black and just a little too tight. The 3" heels show off her shapely legs, giving her calves just a little extra curve. Taking a huge gulp of air and slowly exhaling, she opens the door. She does not let him see her gasp for breath. He stands 6'2", incredible tan, blondish-brown hair, black eyes, his muscles ripple under his light blue, open 3 buttons down, shirt. His shoulders are broad and taper down to his waist where his 501's begin. Oh, those 501's! He grins with the whitest teeth as she takes his hand and leads him to the kitchen. She pours them both a glass of red wine and their eyes meet during that first sip. All of a sudden the room isn't so casual anymore, and it's too warm. She places the baked potatoes, salad, and lobsters on the table. Sitting down, they both look deeply at each other and proceed to eat without words spoken. Steam rising, potatoes being laden with butter and sour cream, lobsters being cracked with their juices flowing over fingers and onto plates. Fingers being licked. And, their eyes never parting. She picks up a piece of lobster, dips it into the butter and slowly offers it to him. Butter is cascading down her arm leaving a shiny silvery stream. His lips accept the offering and he sucks the meat between his lips and with his tongue licks her fingers, one by one. Time is frozen. Nothing exists. They know without words what is to come next. He rises and walks to her. Taking her hand he pulls her swiftly to her feet. They stare at each other for what seems an eternity. Then turning slowly they reach for the tablecloth, and in one quick motion pull it to the floor - and just do it right there on the table!

GLOSSARY

24/7 • 24 hours/7 days a week.

BBW • big, beautiful woman.

BDSM • bondage, discipline, sadism and masochism.

Bondage • restraining a person for sexual/erotic means by using various kinds of restraints.

Boots • footwear - a fetish for boots is commonly associated with domination.

Bottom • a submissive or sub.

CBT • cock and ball torture.

Domination, Dominant, Dom • a person, either male or female, who controls the lifestyle or physical activities of a submissive or bottom partner.

Femdom • a female dominant.

FF • fist fucking.

FMF • female male female.

GBG • great big grin.

GS • golden shower.

H2E • enemas.

Horse • an item that allows easy bondage to for the purpose of flagellation to ones buttocks, i.e. chair, carpenters horse, or specially built equipment. Could also have second meaning—see Pony.

IC • instant chat—used for instant communication between two parties who are on-line.

ISO • in search of. As ISO another swinger.

K9 • sex with dogs.

Leather • popular for SM interests and especially the gay male SM community.

LOL • laughing out loud.

MFM • male female male, or male for male.
NULLO • as in nullification. Castration.

Pony • a submissive (either male or female) being dressed in bridle or harness and saddled or hitched to a cart for the dominants use.

Puke • vomit, regurgitate.

RAW • crude or rough sex. Usually pertaining to the male raunch and pig lifestyle.

Rimmed, rimming • oral sex with anus area.

ROFL • rolling on floor laughing.

ROFLMAO • rolling on floor laughing my ass off.

RT or R/T • pertaining to real time. Not on computer.

Rubber • a popular material for use in the BDSM area, i.e. hoods, clothing, etc.

Sadomasochism • a form of sexual activity that usually involves bondage, pain, and domination.

SCAT • a term for the enjoyment of feces.

Showers • golden showers or brown showers.

Sissy • submissive males who dress and act very feminine. Also referred to as sissy maid.

Slam • rough sex.

SM, S/M, S&M • sadomasochism, or could be slave/master or mistress.

Top • a dominant or dom.

WS • water sports (relating to urine).

Donna Tracy

Worn on Left (top)	HANKY CODE FOR GAY MEN	Worn on Right (bottom)
Worn on Left (top)	COLOR	Worn on Right (bottom)
Pilot/flight attendant	Airforce Blue	Likes flyboys
Two tons of fun	Apricot	Chubby chaser
Rimmer	Beige	Rimmee
Heavy SM top	Black	heavy SM bottom
Has/takes videos	Black Velvet	Will perform for camera
Safe sex top	Black/White Check	Safe sex bottom
Likes black bottoms	Black/White Stripe	Likes black tops
Scat top	Brown	Scat bottom
Headmaster	Brown Corduroy	Student
Uncut	Brown Lace	Likes uncut
Circumsized	Brown Satin	Wants circumsized
Likes latino bottoms	Brown/White Stripe	Wants latino tops
New in town	Calico	Tourists welcome
Latex fetish top	Charcoal	Latex fetish bottom
Bartender	Cocktail Napkin	Bar groupie
foot fetish top	Coral	Foot fetish bottom
comes in scumbags	Cream	Sucks it out
Tit torturer	Dark Pink	Tit torturee
2-handed fister	Dark Red	2-handed fistee
Wears a dirty jock	Dirty Jockstrap	Sucks it clean
Tearoom top (pours)	Doily	Tearoom bottom (drinks)
Bestialist top	Fur	bestialist bottom
Spanker	Fushia	Spankee
Two looking for one	Gold	One looking for two
Likes muscleboy bottoms	Gold Lame	Likes muscleboy tops
Bondage top	Grey	Fit to be tied
Actually owns a suit	Grey Flannel	Likes men in suits

168

Milker	Holstein	Milkee
Likes to nibble	Houndstooth	Will be bitten
Daddy	Hunter Green	Orphan boy looking for daddy
Hustler (for rent)	Kelly Green	John (looking to buy)
Chicken	Kewpie Doll	Chicken hawk
Needs a place to stay	Keys in Back	Looking for a ride
Has a home	Keys in Front	Has a car
Stinks	Kleenex	Sniffs
Likes drag queens	Lavender	Drag queen
Has tattoos	Leopard	Likes tattoos
Dildo fucker	Light Pink	Dildo fuckee
Dines off tricks (food)	Lime Green	Dinner plate
Wants head	LT Blue	Cocksucker
Likes black suckers	LT Blue/Black Dots	Likes to suck blacks
Likes latino suckers	LT Blue/Brown Dots	Likes to suck latinos
Likes white suckers	LT Blue/White Dots	Likes to suck whites
Sailor	LT Blue/White Stripe	Looking for salty seamen
Likes asian suckers	LT Blue/Yellow Dots	Likes to suck asians
Suck my pits	Magenta	Armpit freak
Cuts	Maroon	Bleeds
Into navel worshippers	Mauve	Has a navel fetish
Cop	Medium Blue	Copsucker
Outdoor sex top	Mosquito Netting	Outdoor sex bottom
Hung 8" or more	Mustard	Wants a big one
Fucker	Navy Blue	Fuckee
Military top	Olive Drab	Military bottom
Anything anytime	Orange	Nothing now (cruising)
Wears boxer shorts	Paisley	Likes boxer shorts
Spits	Pale Yellow	Drool crazy
Piercer	Purple	Piercee

Fist fucker	Red	Fist fuckee
Furry bear	Red/Black Stripe	Likes bears
Park sex top	Red/White Gingham	Park sex bottom
Shaver	Red/White Stripe	Shavee
69er	Robin's Egg Blue	Anything but 69
A cowboy	Rust	His horse
Starfucker	Silver Lame	Celebrity
Smokes cigars	Tan	Likes cigars
Cock & ball torturer	Teal Blue	Cock & ball torturee
Cuddler	Teddy Bear	Cuddlee
Bathhouse top	Terrycloth	Bathhouse bottom
Skinhead top	Union Jack	Skinhead bottom
Beat my meat	White	I'll do us both
Likes white bottoms	White Lack	Likes white tops
Voyeur	White Velvet	Will put on a show
Hosting an orgy	White/Multicolor Dots	Looking for an orgy
Pisser/WS	Yellow	Piss freak
Likes asian bottoms	Yellow/White Stripe	Likes asian tops
Has drugs	Zip-loc Baggie	Looking for drugs
rides a motorcycle	Chamois	Likes bikers

APPENDIX

I have tried to present a sampling of web sites that are related and available, free of charge, on the various lifestyles we have explored from the Internet chatrooms. Please note that web sites come and go on the World Wide Web with regular consistency and I cannot guarantee that all of these web addresses will be working in the future.

BDSM LIFESTYLE:

www.queernet.org/deviant - Deviant's Dictionary. Invaluable web site. Contains a dictionary with definitions for the BDSM lifestyle, including hankie and flag codes. This site helped me tremendously in my research.

www.davisworks.dynip.com/sinner—Sinner~n~Saint. Really nice web site for D/S lifestyle. Includes a Dominants View and a Submissives View of the lifestyle.

www.castlerealm.com—The Castle Realm. D/S resource center. Helpful explanations for this lifestyle. Excellent links.

www.nexie.com/wickedmztoni—WickedMzToni's. Caution, graphic photos. MzToni tells of her journey, along with her partner, of becoming a dominatrix by accepting a dare to visit a BDSM chatroom. Good links.

www.drkdesyre.com—BDSM resource site. Click on Organizations for state links to groups and nightclubs for BDSM. Contains a wealth of links for information. Excellent web site.

www.sanctuarydallas.com—Dallas, Texas dungeon club. Nice site with interior photos, and links. Good explanation of how a BDSM club works. Recommended web site.

GAY LIFESTYLE:

www.dungeonmale.com—gay male BDSM web site created by Master Eric who resides in San Francisco. Very informative for the gay male as site includes activities and events in the Bay area.

www.gayscape.com—excellent web site. Contains links to information on the gay community in every state and other countries. Good links for the lesbian community, also.

www.allpaths.com/rainbow—rainbow resources. Lots of links for lesbian and gay information. Contains, businesses, campus life, and politics, also. Very informative site.

COUPLES/SWINGERS LIFESTYLE:

www.etnsc.com—Eastern Tennessee Swingers Club. Terrific site to discover the swinger's lifestyle. Has great links and even a Swinger Test to see if you might qualify for this style of living.

www.freedomacres.com—Freedom Acres, a California swinger's club. Very professional web site with lots of information on swinging. Be sure to take the photo tour. Highly recommend this site if you are interested in this lifestyle.

www.swingstream.com—Good information and links. Be sure to check out their swing links.

MISCELLANEOUS:

www.thedoghouse.org—The DogHouse. Web site for SM Dog Training. Includes Pony links. Photos.

www.sextoy.com—Adult sex toy store. Interesting site, especially for the beginner if you have no idea what kind of sex toys are available for adults.

ABOUT THE AUTHOR

Born and raised in Hollywood, California, Donna Tracy worked for many years as an actress and studio extra in such films as: *My Fair Lady, Camelot,* and *Robin and Seven Hoods.* Her credits also include the Elvis movies, and many television shows. After working as an actress she moved behind the camera as a Programming Assistant, and continued her love of writing by writing Public Service Announcements (PSA's) for children's television.

Having written her entire life, Donna has had writings published in local community magazines and brochures, and for many years was focused on writing poetry of a spiritual nature, which included finding your inner being. Her first non-fiction work, *Chatroom Voyeur,* was compiled over a two-year period after her curiosity got the best of her, and she had to find out what was transpiring in Internet chatrooms.

Donna's personal experience with Internet chatrooms for over six years, led her to meeting a wonderful man, whom she has been married to for three years now. She states, "Don't ever be lonely." "Get on that computer and meet some terrific people—they really are out there!" "Just be sure to use the common sense that God gave you."

Donna is now fast at work on her second non-fiction novel, *Hollywood Extra,* and resides in the south with her husband, Don. Together they enjoy traveling, spending time in Las Vegas, and enjoying their families.